D1700399

PERGAMON INTERNATIONAL LIBRARY
of Science, Technology, Engineering and Social Studies

The 1000-volume original paperback library in aid of education,
industrial training and the enjoyment of leisure

Publisher: Robert Maxwell, M.C.

NUCLEAR FUEL ELEMENTS

Design, Fabrication and Performance

THE PERGAMON TEXTBOOK
INSPECTION COPY SERVICE

An inspection copy of any book published in the Pergamon International Library will
gladly be sent to academic staff without obligation for their consideration for course
adoption or recommendation. Copies may be retained for a period of 60 days from
receipt and returned if not suitable. When a particular title is adopted or recommended
for adoption for class use and the recommendation results in a sale of 12 or more copies,
the inspection copy may be retained with our compliments. The Publishers will be
pleased to receive suggestions for revised editions and new titles to be published in this
important International Library.

Other Pergamon Titles of Related Interest

CHICKEN	Nuclear Power Hazard Control Policy
COMMISSION OF THE EUROPEAN COMMUNITIES	Fusion Technology 1980, 2 volumes
DOLAN	Fusion Research, 2 volumes
EBERT	Radiation Protection Optimization
FENECH	Heat Transfer and Fluid Flow in Nuclear Systems
HUNT	Fission, Fusion and the Energy Crisis, 2nd Edition
JUDD	Fast Breeder Reactors: An Engineering Introduction
LATHROP	Planning for Rare Events: Nuclear Accident Preparedness and Management
LEWINS	Nuclear Reactor Kinetics and Control
MICHAUDON	Nuclear Fission and Neutron-Induced Fission Cross-Sections
MURRAY	Nuclear Energy, 2nd Edition
PETROSY'ANTS	Problems of Nuclear Science and Technology
SHABALIN	Fast Pulsed and Burst Reactors
SILVENNOINEN	Reactor Core Fuel Management

Related Journals Published by Pergamon Press

(Free Specimen Copies Gladly Sent on Request)

Annals of Nuclear Energy
Annals of the ICRP
Energy, the International Journal
Energy Conversion and Management
International Journal of Applied Radiation and Isotopes
International Journal of Heat and Mass Transfer
Letters in Heat and Mass Transfer
Nuclear and Chemical Waste Management
Progress in Nuclear Energy

NUCLEAR
FUEL ELEMENTS

Design, Fabrication and Performance

by

BRIAN R. T. FROST , B.Sc., Ph.D., M.B.A.
Director of the Materials Science Division,
Argonne National Laboratory, Illinois, USA

PERGAMON PRESS

OXFORD · NEW YORK · TORONTO · SYDNEY · PARIS · FRANKFURT

U.K.	Pergamon Press Ltd., Headington Hill Hall, Oxford OX3 0BW, England
U.S.A.	Pergamon Press Inc., Maxwell House, Fairview Park, Elmsford, New York 10523, U.S.A.
CANADA	Pergamon Press Canada Ltd., Suite 104, 150 Consumers Rd., Willowdale, Ontario M2J 1P9, Canada
AUSTRALIA	Pergamon Press (Aust.) Pty. Ltd., P.O. Box 544, Potts Point, N.S.W. 2011, Australia
FRANCE	Pergamon Press SARL, 24 rue des Ecoles, 75240 Paris, Cedex 05, France
FEDERAL REPUBLIC OF GERMANY	Pergamon Press GmbH, 6242 Kronberg-Taunus, Hammerweg 6, Federal Republic of Germany

Copyright © 1982 Brian R T Frost

First edition 1982

Library of Congress Cataloging in Publication Data

Frost, Brian R. T.
Nuclear fuel elements.
(Pergamon international library of science, technology, engineering, and social studies)
1. Nuclear fuel elements. I. Title.
II. Series.
TK9207.F76   1982   621.48'335   81-21131
AACR2

British Library Cataloguing in Publication Data

Frost, Brian R. T.
Nuclear fuel elements.—(Pergamon international library)
1. Nuclear fuels
I. Title
621.48'335   TK9360

ISBN 0-08-020412-0 (Hardcover)
ISBN 0-08-020411-2 (Flexicover)

In order to make this volume available as economically and as rapidly as possible the typescript has been reproduced in its original form. This method unfortunately has its typographical limitations but it is hoped that they in no way distract the reader.

Printed in Great Britain by A. Wheaton & Co. Ltd., Exeter

Foreword

Nuclear power remains one of the few economically viable and safe options for the supply of electric power, despite vigorous attempts to discredit it. The developed and developing countries are forging ahead with nuclear power plant construction: these are mostly American-designed light water reactors. The fuel cycle, which is the key to the economic success of these plants, continues to attract great attention. The issues of nuclear proliferation and of reprocessing have given increased impetus to raising the burn-up limits of fuel elements and to developing an ability to operate with leaking fuel pins. The Three Mile Island Unit-2 incident has produced an increased sensitivity to reactor safety issues, especially to the behavior of fuel elements in a loss of cooling accident (LOCA).

The continued growth of nuclear energy combined with a greater attention to institutional problems, like reactor licensing, has created a growing demand for properly trained scientists and engineers. Their proper training must include knowledge of fuel element design and performance. The training in this subject has not been as good as it could be. Most of the training tends to be of an apprenticeship nature, i.e. learning while working with experienced practitioners, rather than academic. Yet some academic grounding gives the practitioners great assistance in their task of training their apprentices. One of the difficulties encountered in academic training in fuel element technology has been a lack of up-to-date texts. In 1976 Don Olander published an excellent text ("Fundamental Aspects of Nuclear Reactor Fuel Elements" published by ERDA through the NTIS). This book concentrates on fast reactor fuel elements but contains a lot of fundamental nuclear materials science. In 1981 J. T. Adrian Roberts published his book, "Structural Materials in Nuclear Power Systems" through Plenum Press. This deals with all aspects of reactor materials, including chapters on LWR and LMFBR cores, and is a welcome and valuable addition to the literature.

In this text I have tried to contribute to the teaching and understanding of fuel element technology for all types of reactor by writing a relatively brief "road map" for the development of fuel elements. It is written at a level that may disappoint some of my colleagues who see the need for an authoritative advanced text which would serve the needs of the practitioners. The makings of such exist in a report that David Okrent and several of us at Argonne put together in 1972. To bring that up to date and to edit it properly is an enormous task which hopefully will be done one day. In the meantime I have attempted a more modest effort to write at the graduate student level a text which will introduce the reader to the subject and provide, through references, a means of building on that knowledge.

Fuel element technology is a multi-disciplinary subject, embracing physics, chemistry, heat transfer and fluid flow, applied mechanics, metallurgy, ceramics and systems analysis. The development of analytical modelling techniques in the 1970's has provided an admirable vehicle for bringing these disciplines together, allocating priorities and testing their validity by comparison with real experiments. This development is given some emphasis in this book but to achieve a full understanding the reader must experiment with models for himself with the help of the National Energy Software Center at Argonne which can provide software and instructions.

This book has been a long time in gestation and consequently is very different from the original concept. My thinking has been modified by events, politics and opportunities to teach reactor materials science to nuclear engineers both in England and in America. It is to be hoped that the principles and philosophies in this field will remain unchanged for long enough for this book to fulfill its primary goal of contributing to the training of fuel element technologists so that they may more effectively contribute to this challenging and important field.

In conclusion I would like to acknowledge the help of my colleagues working in this field, Regina Wanda for typing the text, my wife for her encouragement and the Illinois winter weather that persuaded me to stay indoors and write.

Argonne, Illinois   B. R. T. Frost
December 1981

Contents

Chapter 1   Introduction and Philosophy   1
              The reactor operator's approach   6
              The materials scientist's approach   7
              The interdisciplinary approach   10

Chapter 2   Fuel Types   13
              Fuel origins   13
              Fuel forms   18
              Metal fuels   18
              Metallic dispersion fuels   23
              Liquid fuels   23
              Fuel characterization   40
              Appendix: Phase diagrams   41

Chapter 3   Irradiation Behaviour of Fuels   50
              Oxides   57
              Carbides and nitrides   70
              Dispersion fuels   72
              Appendix: Burn-up and rating of nuclear fuels   75

Chapter 4   Cladding and Duct Materials   77
              Design considerations   77
              Commonly used materials and their properties   80
              Compatibility   85
              Radiation effects   88

Chapter 5   Fuel Element Design and Modelling   98
              Some sources of data and codes   130

Chapter 6   Fuel Element Performance Testing and Qualification   132
              Strategies   135
              Post-test examination   138
              Commercialization   140
              Appendix: Suppliers of specialized equipment   142

Chapter 7   Experimental Techniques and Equipment   149
                 Out-of-pile   149
                 In-pile   151
                 Post-irradiation examination   158

Chapter 8   Water Reactor Fuel Performance   177
                 PWR   177
                 BWR   181
                 HWR   181
                 Failure   181
                 Failure mechanisms   184
                 Licensing and regulation   190
                 17 × 17 Fuel surveillance programme   191

Chapter 9   Gas-cooled Reactor Fuel Elements   200
                 Magnox   201
                 French gas-graphite   204
                 AGR   206
                 HTGR   210
                 Safety   216

Chapter 10   Fast Reactor Fuel Elements   217
                 Introduction   217
                 Early reactors   219
                 Prototype reactors   232

Chapter 11   Research and Test Reactor Fuel Elements   247

Chapter 12   Unconventional Fuel Elements   265

Some Useful General References   269
Tabulation of Industrial Capabilities in the USA   271
Index   273

CHAPTER 1

Introduction and Philosophy

A nuclear reactor is basically a heat source in which energy is released
through the fission of an isotope of uranium or plutonium. A fissionable
atom, e.g. ^{235}U, located on a crystal lattice in, say, a pellet made of
sintered UO_2 crystals undergoes fission when a neutron of suitable energy is
absorbed. This occurs because the resulting nucleus is very unstable and
splits into two parts of roughly equal mass, e.g.

$$^{235}U + n^1 \rightarrow {}^{95}X + {}^{139}Y + 2n^1.$$

There is a discrepancy in mass between the two sides of this equation which
corresponds to an energy release of about 200 MeV.[*] Most of this energy is
imparted to the two fission fragments, which leave the site of fission very
rapidly, travelling in straight lines of length about 10 μm before coming to
rest. In doing so they impart their energy to the parent lattice, essentially
as thermal vibrations, and incidentally causing considerable damage to the
lattice. It is this energy, plus a smaller amount arising from other sources,
which represents the heat source in the nuclear fuel which must be converted
to a more useful form.

The fuel element is the fundamental building block of the reactor core. It
contains a discrete quantity of nuclear fuel. While there is a class of
homogeneous reactors in which the fuel is dissolved or dispersed in a fluid
(that may also act as the coolant), the more common practice is to segregate
the solid fuel inside metallic cladding. The most common geometry for these
fuel elements is a long cylindrical rod (Fig. 1.1), although plates are
sometimes used, e.g. in research reactors. The coolant flows under forced
convection over the cladding to remove the heat (cool the fuel element) and
transports it to a heat exchanger (Fig. 2). Different types of reactor call
for different levels of performance from the fuel elements, and this leads to
different designs and choices of materials. For example, a research reactor
has to generate a high flux, but need not generate high coolant temperatures.
In this case the high specific power and surface heat fluxes are best handled
by a plate geometry with a high surface area : volume ratio and by an
aluminium-based fuel which has good neutronic properties but a low melting
point (Fig. 1.3).

---

[*]1 MeV is approximately equal to 1.602×10^{13}J.

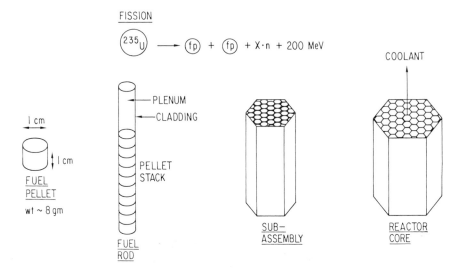

Fig. 1.1.   The building blocks of a nuclear reactor fuel
element, showing the basic nuclear reaction
occurring in the ^{235}U in the pellet which is
loaded into the fuel rod, which is in turn
loaded into a sub-assembly for stacking in the
core.

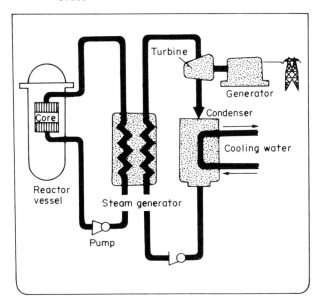

Fig. 1.2.   The heat transport system of a pressurized
water reactor.   Source: A Guidebook to Nuclear
Reactors by Anthony V. Nero, Jr., University
of California Press, 1979.

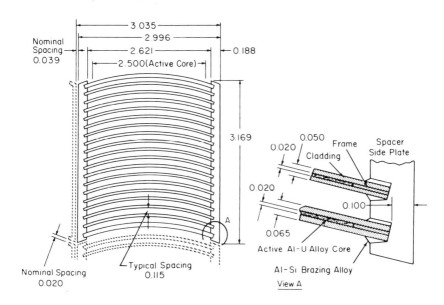

Fig. 1.3.   End view and detail of U-Al plate-type fuel
element for a research reactor.   Courtesy: ORNL.

The starting point for fuel element design and development is the specification
of the purpose and performance requirements of the reactor. The design of the
reactor core involves reactor physics, engineering (heat transfer, fluid flow
and stress analysis), materials science, safety and economics inputs. The
finished product represents a compromise or optimization between these many
different factors. The core and hence the fuel element are designed, built,
operated and refined in an iterative process.

The reactor core is made up of an assembly of fuel elements. Early designs
used individual fuel rods, but it is now standard practice to group many rods
together into a bundle or *subassembly* which is usually enclosed by a metallic
box or *duct* that acts as a flow separator and as a structural member
(Fig. 1.4). Pressurized water reactors are an exception in that their fuel
elements are not enclosed by ducts. The fuel rods are spaced evenly in the
subassembly by means of wire spacers wrapped around them or by metal grids
placed at intervals down the subassembly. A subassembly may generate several
megawatts of heat. It is a convenient sized unit for moving fuel in and out
of the core.

It must be remembered that the fuel element in the reactor core is the middle
stage (albeit the most important one) in the fuel cycle. Fuel is mined,
concentrated, perhaps enriched, fabricated and enclosed, irradiated in the
core, and reprocessed for removal of the fission products which must be
converted to a chemically stable form and stored. These many stages impact
on one another in many ways. However, the purpose of this book is to
describe the materials used in fuel elements and the design, performance and

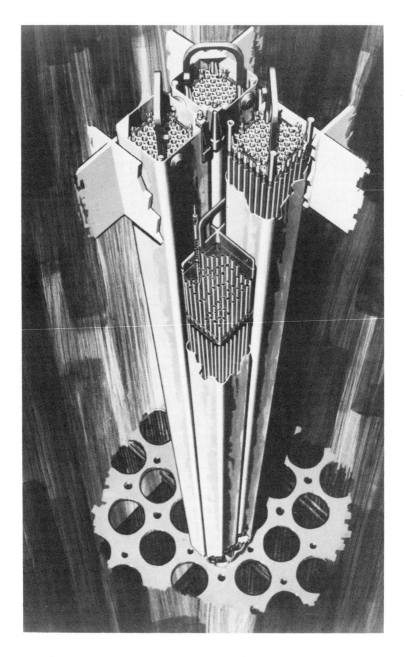

Fig. 1.4.   Fuel element subassemblies for a Boiling Water
Reactor.   Courtesy: General Electric Company.

analysis of fuel elements for a variety of reactor types. Other works deal
with the fuel cycle in its entirety [1].*

The phenomena that occur in a fuel element when it is at power in a reactor
core are numerous and complex. A full understanding of how a fuel element
works involves a knowledge of the fission process, its chemical and nuclear
particle products, their chemical and physical effects on the fuel and
cladding, the heat and mass transport processes that occur as a consequence
of the fission heating, and the chemical reactions that occur between the
coolant and the fuel and the cladding materials. To understand these we must
use our full range of knowledge of chemical thermodynamics and kinetics, of
physical processes such as atomic diffusion and defect behaviour, and of the
engineering sciences of heat transfer and stress analysis. The physical and
chemical states in the fuel rod are never constant; even if the fission rate
is constant, the fission products are building up, moving around and
agglomerating while the atomic defects in the cladding are moving around and
clustering or annihilating themselves, or changing the rate at which physical
processes such as diffusion and creep occur [2,3].

The environment in the reactor core varies from point to point. The fission
rate peaks at the core centre and drops off all around because of neutron
leakage (Fig. 1.5). Generally the coolant flows upward through the core so
that there is a temperature rise from the bottom to the top of the core.
Since the phenomena in the fuel and cladding materials are sensitive to
temperature and to fission rate or neutron flux, each point in the core will
behave differently from the others. This makes fuel element design and
development particularly difficult.

Uncontained radioactivity may endanger the health of humans and animals. The
main consideration in licensing a reactor core, i.e. a fuel element design,
is the assurance that the fuel elements will contain most, if not all, of the
fission products and actinide elements. The cladding is the first line of
defence. Hence we must know as much as possible about the arrangement of the
fission products in the fuel, the integrity and performance limits of the
cladding, and the consequences of a hole or crack forming across the cladding
wall. Let us start analysing this complex situation by viewing it from three
different viewpoints.

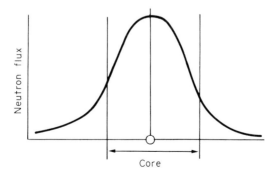

Fig. 1.5.   The variation in neutron flux with core position
(two dimensions). Source: ANL.

---

*Numbers in square brackets refer to References at end of chapter.

THE REACTOR OPERATOR'S APPROACH

Ideally a fuel element should be cheap, reliable and safe. Nuclear reactors have high capital (construction) costs and it is the low fuel cycle cost which allows them to compete with other energy sources. Hence we must aim at a fuel element that can be massproduced with very reproducible performance character- istics and at a reasonable cost. These fuel elements must either have a very predictable life-time before they fail *or* failures must not cause a reactor shutdown, since downtime is very expensive; in 1980 it amounted to about $500,000 per day for an American 1000-MWe reactor. Fuel element failures plotted as a function of time at power or burn-up generally follow a U-shaped distribution and it is the size of the tail on the low end that concerns us (Fig. 1.6).

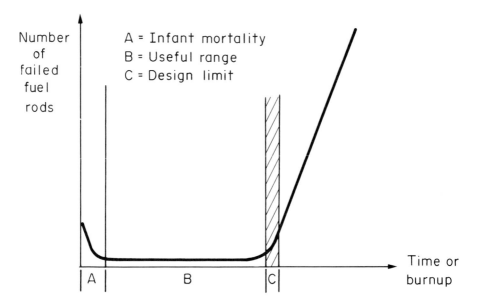

Fig. 1.6. The failure incidence of fuel elements as a function of time in the core. Source: ANL.

All reactors are subject to power cycles, if only on start-up and shutdown, and may occasionally experience a mild overpower excursion. The fuel elements must be able to withstand this cycling without an increased susceptibility to failure. Failure in this context means a breach of the cladding, such as the appearance of a pinhole or a small crack in the cladding whose occurrence should not lead to progressive failure. To illustrate this point, uranium metal when exposed to water oxidizes, the reaction product causing progressive opening of the initial hole or crack. On the other hand, UO_2 reacts very slowly with water and a UO_2-Zircaloy fuel element in a commercial light-water reactor is usually allowed to remain in the core until a scheduled shutdown.

Fuel elements should be designed to behave in the most favourable manner in the unlikely event of a major malfunction, such as a loss of coolant flow or an overpower type of accident. In such accidents the cladding may melt and the fuel may partially or totally melt. We must attempt to understand the details of such processes and to minimize the consequences. Attempts have been made to design fail-safe elements, especially for fast reactors. Ideas which have been examined are fusible links to drop blocked elements out of the core and the use of annular elements in the Dounreay Fast Reactor* with different cladding materials on the two annuli, the inner of which reacts with the fuel and melts first (see Chapter 10).

A more subtle effect that had safety and licensing implications is the change in UO_2 fuel density that can occur early in life. This densification problem initially caused the collapse of cladding and unplanned neutron streaming in LWRs. A simple expedient of prepressurizing the pins with helium stopped that effect and a more lengthy study of the densification process itself led to the development of a sintered UO_2 pellet that does not densify. This is discussed in more detail later in Chapter 8.

Generally the reactor owner-operator does not design and produce his own fuel elements. He writes a performance specification, an important part of which is the definition of the target burn-up. The fuel vendors quote a price to meet the specification and the lowest bidder supplies the fuel, usually with a guaranteed burn-up. Fuel elements that fail prematurely are replaced by the vendor. Given the great sensitivity of reactor economics to the integrity of the fuel elements, this is probably the only viable system.

THE MATERIALS SCIENTIST'S APPROACH

When uranium or plutonium atoms fission, they produce a range of fission products which are depicted in Figs. 1.7 and 1.8. These elements have a large influence on fuel element behaviour. The noble gases xenon and krypton are produced in fair abundance, including by decay of precursors such as iodine. These gases can behave in a variety of ways that we will discuss in detail later; however, we can broadly think in terms of (1) the gases remaining in the fuel and joining together to form bubbles which cause the fuel to *swell*, or (2) the gases diffusing to the surface of the fuel and being *released* into the sealed cladding, building up a pressure. In either case the cladding is subjected to a steadily rising pressure, which will eventually exceed the external coolant pressure and cause the cladding to fail. Hence fission gas behaviour is a strong determinant of fuel element design and performance.

Some of the more volatile fission products, such as iodine and caesium, migrate to the cooler regions of the fuel element, i.e. the cladding, where they may induce stress-corrosion cracking or other degradation phenomena. The discontinuities in the fuel surfaces due to pellet-pellet interfaces and to cracks result in locally high stresses on the cladding and are favoured sites for chemical reactions by fission products. The term "pellet-cladding interactions", or PCI, is used to describe these effects in LWR fuel elements that are mainly responsible for early failures. Fuel element designers have attempted to overcome this problem in a variety of ingenious ways as discussed in a later chapter.

---

*This reactor has now been decommissioned. Its fuel element is discussed in more detail in Chapter 10.

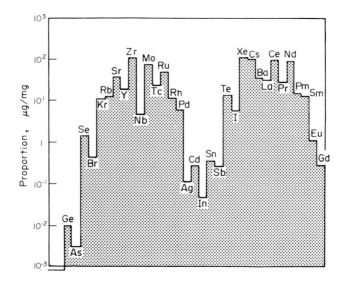

Fig. 1.7.   The mass distribution of fission products from
thermal fission of ^{235}U.   Source: B. R. T. Frost,
R.I.C. Reviews, 1969.

The cladding is the only barrier to fission products entering the coolant
stream, i.e. it is the primary containment for the fission products and it
must remain intact throughout its arduous lifetime. Cladding is restricted
in thickness by physics and thermal stress considerations which are discussed
later. It must, however, be thick enough and strong enough to resist
swelling, coolant and fission gas pressures. The latter can be reduced by
leaving an empty space or "plenum" above or below the fuel, provided the fuel
element is of the full core length (see Fig. 1.1). A stack of short elements,
as used in the CANDU and AGR reactors, cannot have a plenum and must be
designed to limit fission gas release.

Provided that chemical effects can be overcome, the cladding will fail when
its ductility is exhausted. Most cladding materials are normally quite
ductile. However, radiation effects reduce the ductility very considerably
— crudely it can be related to the work-hardening of metals. In a fast
neutron flux voids or low-pressure bubbles form and cause the cladding to
swell. This phenomenon causes considerable difficulties in fast reactor core
design because the two opposite faces of a hexagonal subassembly duct are
exposed to differing fluxes. This causes differential swelling and hence
bowing, and interference stresses which can damage the fuel. Thus there is
a high premium on developing duct materials with low swelling rates. The
swelling of the cladding, if uniform, may be advantageous in that it reduces
or totally removes the fuel-swelling stresses on the cladding because the
latter is always moving away from the fuel.

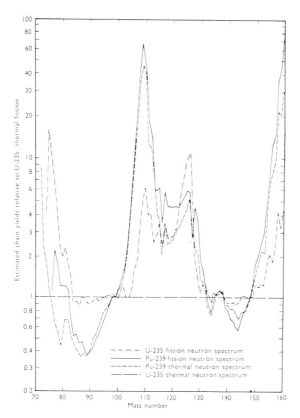

Fig. 1.8. The relative fission yields of ^{235}U and ^{239}Pu.
Source: B. R. T. Frost, R.I.C. Reviews, 1969.

One can easily see that the materials scientist's approach to fuel element
design and development is both rigorous and complex. It is uncommon in
engineering practice to approach a design problem with the rigour that is
applied to fuel elements. The traditions of the field are such that a great
deal of fundamental research has been carried out on the various phenomena
that occur in fuel elements — especially on radiation damage, fission product
behaviour and fuel properties. Given the seriousness with which fission
product release is regarded by the regulatory commissions and the public, it
is necessary to apply all of this knowledge in order to understand and predict
fuel element behaviour. One way in which the materials scientist does this,
which also helps the reactor designer, is to develop computer models of fuel
element behaviour [4]. These codes are usually based on physical models of
materials behaviour, although some use empirical relations. The physically-
based codes require a lot of reliable materials input data before they can
be expected to give realistic results. Conversely, these codes have proved
to be very useful in defining which property data are essential to fuel
element design and hence in guiding the supporting research. Models and
codes are used extensively in reactor licensing activities and for safety
analyses where the experiments are often very complex and expensive, although
the final test of any code is its checkout against well-controlled experiments.

THE INTERDISCIPLINARY APPROACH

The development of a fuel element is a lengthy, iterative process. With 35 years of reactor technology behind us it is rare that a fuel element is now designed from scratch. More frequently it is a case of modifying existing designs; for example, in the case of pressurized water reactors the initial fuel element design was developed in the early 1950s for naval reactors. The first commercial plants, such as Indian Point I, were designed on the basis of the naval reactor experience. These reactors were fairly small and the elements relatively short. From these elements were developed the longer and bigger PWR elements and subassemblies for 1000-MWe plants. Today these elements are being modified and tested to achieve higher burn-ups in support of the proliferation-resistant once-through cycle.

The initial phase of the development of a fuel element is the interaction between reactor physicists, engineers, materials scientists and safety experts in which a goal is specified and then a concept is developed through a series of compromises or trade-offs (see Table 1.1). Once a concept is specified, experimental work is started to:

1. Perform physics calculations and verify these through critical assembly experiments.
2. Better define the thermal-hydraulics parameters through rig and loop tests.
3. Set up small-scale fabrication facilities for fuel, cladding and fuel element assembly.
4. Measure fuel and cladding properties out-of-pile, e.g. fuel thermal conductivity and fuel-cladding compatibility.
5. Conduct small-scale experimental irradiation experiments in research or test reactors; some may be in special loops.
6. Begin to develop a computer model of the concept.

This phase is likely to take at least 3 or 4 years, and as a result the concept is likely to be modified and improved. The next phase is to fabricate and irradiate small numbers of fuel assemblies of this improved design in test reactors and/or in commercial reactors to get limited statistical data and to provide irradiated elements for off-normal and safety tests which will follow the safety analyses conducted at this time. It will probably be necessary or desirable to conduct carefully instrumented in-pile experiments to verify the design calculations and out-of-pile data. These may be "separate effects" tests to verify single parameters or "multiple effects" tests to verify a combination of parameters.

The final phase is either the construction of a new reactor, incorporating a full core loading of this design, or a partial core loading in a "commercial" reactor. It is common for successive core loadings to differ, i.e. there is a continual process of upgrading the design, as long as this is consistent with fabrication economics.

It should be emphasized that most of this research and development is very expensive. It will involve special facilities such as glovebox lines, fabrication plants with rigid quality assurance, criticality and safeguards controls, reactor operations costing millions of dollars per year, and hot cells capable of sophisticated examinations at a radiation level of 10^5-10^6 Ci of γ activity plus α and β. Hence, all of the worldwide reactor research and development up to commercialization has been funded by national governments. The scale of the LWR fuel business is now large enough for commercial vendors to fund much of the upgrading work. However, with the advent of the energy crisis and nonproliferation policies, more government and utility money is going towards fuel element improvement.

To summarize; this book is intended to be a "road map" to the design, development and use of nuclear fuel elements. While this is an inter-disciplinary activity, considerable emphasis will be given to the materials science aspects because further progress will only be made through a thorough understanding of the phenomena that limit fuel element performance. We will review fuel and cladding separately and then combine them in a key chapter on design. We will then discuss the facilities that are needed to proceed from a design to commercial use, and finally we will review experience with fuel elements in most of the common reactor types.

TABLE 1.1.   Stages in Fuel Element Development

Feedbacks

1. Define the reactor type: purpose, coolant, performance.

2. Calculate fuel element dimensions, heat fluxes, design features, physics, critical assemblies.

3. Consider the range of available fuels and cladding materials and pick the most suitable for the design.

4. Consider fuel and cladding properties — out-of-reactor
   — in-reactor

5. Examine fuel-cladding interactions (chemical and physical).

6. Develop a preliminary fuel element design.

7. Develop fabrication procedures for fuel, cladding and fuel element if necessary.

8. Develop mathematical models of the fuel element — specify and obtain input data.

9. Test fuel elements — out-of-reactor: thermal-hydraulics
   — in-reactor: on increasing scale
   — transient tests

10. Analyse tests — post-irradiation examination
    — failure mechanisms
    — run beyond cladding breach
    — feedback to models
    — improvements in design

11. Optimization via iterations of 8, 9 and 10. Write detailed design and manufacturing specifications.

12. Production — quality assurance, NDE, SPM assay
    — safeguards, safety, criticality control
    — economics, automation

13. Establish interfaces with the rest of the fuel cycle — mining, enrichment, reprocessing, waste management.

14. Licensing of qualified cores.

REFERENCES

1. Elliott, D. M. and Weaver, L. E. (eds.), *Education and Research in the Nuclear Fuel Cycle*, University of Oklahoma Press, Norman, OK (1972).
2. Frost, B. R. T., *Royal Institute of Chemistry Reviews*, 2(2), 163-205 (August 1969).
3. Olander, D. R., *Fundamental Aspects of Nuclear Reactor Fuel Elements*, TID-26711, ERDA (1976).
4. Harris, J. E. and Sykes, E. C., *Physical Metallurgy of Reactor Fuel Elements*, Session 6: Fuel-element performance modeling, The Metals Society, London (1975).

CHAPTER 2

Fuel Types

FUEL ORIGINS

Only one nuclear fuel occurs naturally: uranium, which contains only 0.7% of the fissionable isotope ^{235}U. The more abundant isotope ^{238}U may be converted to plutonium (^{239}Pu) by a neutron capture and decay process (Table 2.1), and ^{239}Pu is readily fissionable. Thorium occurs abundantly and forms fissionable ^{233}U through a neutron capture and decay process (see Table 2.1). Hence there is one "basic" fuel — uranium — and two other elements — thorium and plutonium — which enter the fuel cycle through breeding processes.

TABLE 2.1. Principal Fissionable Nuclides

Nuclide	Fission threshold (neutron energy in MeV)	Average energy available from thermal fission (MeV)
Fertile ^{232}Th	1.4	—
Fissile ^{233}U	0	198
Fissile ^{235}U	0	202
Fertile ^{238}U	0.6	—
Fissile ^{239}Pu	0	210
"Uranium–plutonium cycle"	$^{238}U + n \xrightarrow{\gamma} {}^{239}U \xrightarrow{\beta} {}^{239}Np \xrightarrow{\beta} {}^{239}Pu$	
"Thorium–uranium cycle"	$^{232}Th + n \xrightarrow{\gamma} {}^{233}Th \xrightarrow{\beta} {}^{233}Pa \xrightarrow{\beta} {}^{233}U$	

Source: Anthony V. Nero, *A Guidebook to Nuclear Reactors*, University of California Press (1979).

Uranium and plutonium can be used in a number of different forms as nuclear fuels: generally as metals and alloys or as ceramic compounds such as oxides and carbides. These fuel forms may be introduced into the fuel element in bulk, e.g. cast metal rods or sintered oxide pellets, or they may be dispersed in a nonfissionable metal or ceramic matrix. These options will be discussed later in relation to fuel element performance. At this juncture one needs to keep in mind the fuel form that is to be made for loading into a reactor, while the route from the uranium mine (the fuel cycle, see Fig. 2.1) is briefly considered [1].

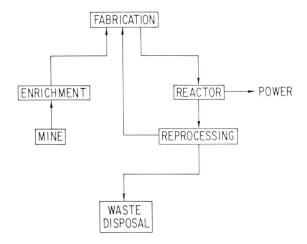

Fig. 2.1.   A schematic representation of the nuclear fuel
cycle.   Source: Author.

Uranium is generally mined as uraninite, a form of pitchblende which contains
a mixture of uranium oxides (mostly U_3O_8) at a concentration of 0.1% or
higher.   The ore is crushed, concentrated and the uranium is leached into
acid or sodium carbonate.   The crude leachate is purified and converted to
pure uranyl nitrate which is in turn converted to UO_2 (see Fig. 2.2).   The
way in which this conversion process is carried out very significantly
influences the properties of the UO_2 powder which, in turn, influence the
properties of the sintered powder.

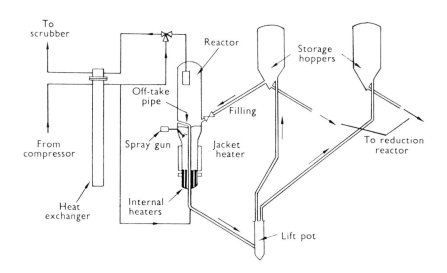

Fig. 2.2.   Schematic diagram of the process for the
denitration of uranyl sulfate.   Source:
B. R. T. Frost, R.I.C. Reviews, 1969.

Generally uranium is enriched by gaseous diffusion or centrifuge cascade processes in order to increase the ratio of ^{235}U : ^{238}U to satisfy reactor physics requirements. In this event UO_2 is converted to UF_6 for enrichment and subsequently to UF_4 if metallic uranium is required, or to UO_2 if a ceramic is required. The fluorination process can introduce undesirable impurities into the final fuel form unless considerable care is taken.

Metallic uranium is produced by reacting UF_4 with high-purity calcium or magnesium in a sealed "bomb", the heavy metal solidifying as a large nugget or "derby" in the base of the bomb (Fig. 2.3). The metal is then cleaned, remelted under vacuum in an RF furnace and cast into rods close in size to the final fuel rod. This last stage may include the addition of alloying elements to improve irradiation performance or corrosion resistance. A great deal of detailed information on uranium metallurgy is given in the book *Uranium* by Gittus [2].

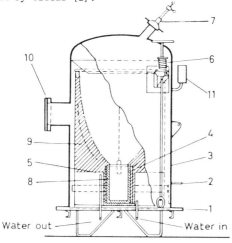

1—base plate, 2—exhaust tube and argon inlet, 3—bell, 4—steel crucible, 5—interior of furnace 6—closing device, 7—firing device, 8—crucible of fritted fluorite, 9—lining of ram-packed fluorite, 10—side opening, 11—manometer.

Fig. 2.3.   Apparatus for making 80 kg uranium ingots by calcium reduction of UF_4. Source: Uranium, J. H. Gittus, Butterworths, 1963.

UO_2 powder may be consolidated into dense pellets, or it may be converted to other compounds, such as UC and UN. For example, UC is most commonly made by mixing UO_2 powder with carbon and heating to ~1500°C in a vacuum or in a rapidly flowing inert gas. This produces a powder which is subsequently consolidated into dense pellets. As will be discussed later, dispersion fuels require the conversion of UO_2 or UC powder to small spheres for incorporation into a metal or ceramic matrix and this requires special considerations.

In all stages of fuel fabrication great attention has to be paid to quality control, toxicity, criticality hazards, and safeguards against diversion. Quality control begins with uranyl nitrate and its conversion to UO_2 or UF_6 and continues through to the loading of the fuel elements into a reactor. A very large number of pellets or even fuel rods is required to load a power reactor core;[*] consistency and predictability of performance are extremely important. Hence fuel (and cladding) specifications are carefully written and are enforced through rigid quality assurance procedures that are comparable to those used in the food, drug and aerospace industries.

Uranium is only mildly toxic, although fine UO_2 (and UC) powder must be handled with care. Plutonium is highly toxic. Hence, all operations involving plutonium and PuO_2 powders are conducted inside gloveboxes which are under a negative pressure of inert gas. This means, of course, that all operations are slower and more awkward — or they must be fully automated. Glovebox operations are discussed more fully in Chapter 6.

When the ^{235}U, ^{233}U or ^{239}Pu content of fuel is above about 15%, regulations must be enforced to ensure that critical assemblies are not formed. Critical quantities in a fully moderated state are calculated and quantities of material in process are carefully controlled by administrative procedures and through careful design of the dimensions of process equipment. Neutron detectors are strategically placed to warn of approach to criticality.

There has always been responsible concern in the nuclear industry for the control of access to fissile material. This has been intensified in recent years through the emergence of organized terrorist groups and a perceived desire of certain groups to attain a nuclear weapons capability. Safeguards of nuclear materials are exercised through a variety of procedures that include careful security checks of all personnel, fences, guards, dogs, metal detectors, etc. New antiproliferation methods include adding strong gamma-emitting materials to the fuel to discourage theft.

While these many procedures make it more difficult to develop and produce nuclear fuels, they do not affect the rate of progress and the economics very significantly when applied intelligently and realistically (as opposed to bureaucratically). It is not intended to elaborate on this aspect of fuel element development in this book. However, these difficulties must be faced when considering this subject.

To complete this brief discussion of the fuel cycle, we must consider the fate of the fuel elements after they have reached the end of their useful life in the reactor core. They are, of course, removed from the core and placed in storage pools at the reactor site. To complete the fuel cycle they should be shipped to a reprocessing plant where the cladding is removed and the fuel is dissolved in acid. The solution is passed through columns where it contacts organic solvents, such as tributylphosphate, which separate out the uranium, plutonium and fission products. The uranium and plutonium are returned to the fuel fabrication plants while the fission products are disposed of. Up to the present the waste products have been stored in solution in large tanks. A serious effort is now being made to convert the waste to a solid, stable, insoluble form. The favoured route is to mix the fission product oxides with silica sand and heat the mixture to form a glass which is encapsulated in metal and buried in a geologically stable medium, such as a salt mine.

In 1981 reprocessing is at a standstill in the USA because of the belief that the separation of plutonium may lead to its diversion to make a weapon. Walter Marshall of the UK, on the other hand, has stated that the safest

---

[*]An 1100-MWe pressurized water reactor has about 50,000 fuel rods in the core, containing 12 million UO_2 pellets.

place for plutonium is in a reactor rather than in a spent fuel storage pool, where they rest today, leading to a serious problem of what to do when the storage pools are all full.

On the other hand, efforts on waste storage have increased considerably and several countries have built pilot-scale plants to convert waste to glass.

The complexity of the total fuel cycle is illustrated in Fig. 2.4 in which data are given for the quantities of materials involved in the total fuel cycle for a 1000-MWe plant.

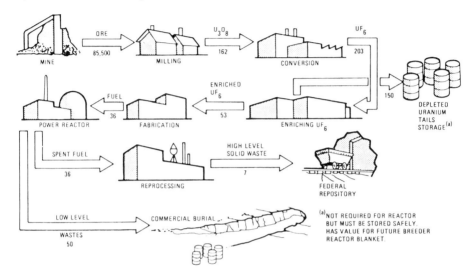

Fig. 2.4. Annual quantities (tons) of fuel materials required for routine (equilibrium) operation of 1,000 MW(e) light water reactor.
Source: ERDA.

The subject of fuel cycle costs is a complex one that will not be discussed at length here. However, we should note that the high capital cost of nuclear plants must be offset by low fuel cycle costs in order to produce electricity at a price that is competitive with coal. This is a rapidly changing scene as on the one hand the nuclear fuel cycle costs are increasing due to increased safeguards, increased ore costs and the cost of waste management, while coal costs are increasing due to the need to transport low-sulphur coals over long distances, the cost of stack-gas scrubbers and scrubber waste disposal, and the rising cost of coal at the mine.

Mason [1] has described the cost of electricity e in terms of a number of parameters:

$$e = 1000 \frac{\phi I + O + F}{E} \text{ mills/kWh(e)} \tag{2.1}$$

where ϕ = annual fixed charge rate, yr^{-1},
  I = initial cost of plant, \$,
  O = annual operating cost, \$/yr,
  F = annual fuel cost, \$/yr
  E = net electricity generated, kWh(e) yr^{-1}.

Since $$E = 8760LK = 24\eta BU \tag{2.2}$$

where L = capacity factor,
K = rated net plant capacity,
B = fuel burn-up at discharge MWD/teU,
U = nuclear fuel consumption kg yr^{-1},
η = plant thermal efficiency,
8670 = number of hours/yr,

$$F = C_F U \tag{2.3}$$

where C_F = total fuel cycle costs/kg uranium.

Thus we can rewrite equation (2.1) as

$$e = \frac{1000}{8670L} \underbrace{\left(\phi \frac{I}{K} + \frac{O}{K} \right)}_{\substack{\text{fixed} \\ \text{costs}}} + \underbrace{\frac{1000}{24} \cdot \frac{C_F}{\eta B}}_{\text{fuel costs}} \cdot \tag{2.4}$$

Note particularly that electricity costs vary inversely with fuel burn-up. We can break C_F down into the cost of ore, conversion and enrichment, fabrication, spent fuel shipping and reprocessing and waste management from which must be subtracted the credit for the uranium and plutonium recovered in reprocessing.

FUEL FORMS

Nuclear fuels may be grouped into three different classes: metallic, ceramic and dispersions. The latter includes two-phase metallic fuels such as Al-UAl$_3$ as well as dispersions of ceramics in metal or ceramic matrices. We will discuss how these fuels are fabricated, their properties and their behaviour in and out of a reactor and then make some comparisons between them. Table 2.4 summarizes the important properties of the fuels that are discussed below.

METAL FUELS

The most dense form of uranium is the metal; hence it is the form that has been used where enrichment is unavailable or undesirable. A great deal of experience was gained in the use of metallic uranium in reactors built to produce weapons-grade plutonium [3].

Uranium metal melts at 1130°C and has three crystalline forms in the solid state. α-U, stable up to 661°C, has an orthorhombic crystal structure (Fig. 2.5), β-U is stable between 661° and 769°C and has a tetragonal structure, while γ-U is stable between 769°C and the melting point (1130°C) and has a body-centred cubic structure. Normal commercial quality uranium has a purity of 99.9%, with 0.1% of impurities distributed as shown in Table 2.2. Hence the as-cast structure is strongly cored (non-uniform in composition) and the alpha grains have some preferred orientation or texture. The metal is fairly easily worked and the structure can be homogenized. However, cycling the material in the alpha phase at elevated temperatures causes cyclic growth. The orthorhombic crystals elongate in the a-direction and shrink in the o-direction (see Fig. 2.5). Mayfield [4] found that the cyclic growth increases the higher the temperature to which the metal is raised within the alpha range (Fig. 2.6). In a totally randomly oriented piece of uranium these effects cancel out (although internal strain is produced). However, most pieces have some preferred orientation. Clearly this must be eliminated for in-reactor use since a fuel rod must undergo numerous power cycles and hence thermal cycles.

TABLE 2.2. A Typical Analysis for a Cast Uranium Ingot

Test	Typical analysis
Density	18.96 g/cm^3
	ppm
Carbon	400
Hydrogen	1
Chloride	5
Silica	50
Nitrogen	50
Iron	50
Manganese	13
Boron	< 0.2
Cadmium	< 0.2
Chromium	20
Magnesium	5
Silver	< 1
Nickel	40

Source: J. H. Gittus, *Uranium*, Butterworths (1963).

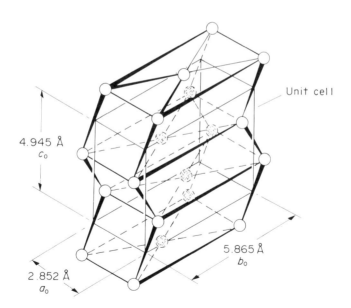

Fig. 2.5. Crystal structure of alpha-uranium.
Source: Uranium, J. H. Gittus, Butterworths, 1963.

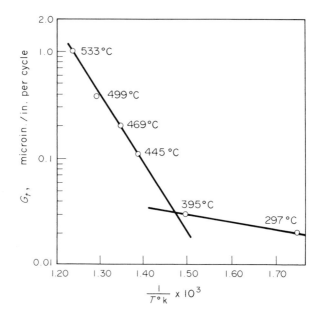

Fig. 2.6.   Growth rate versus reciprocal temperature for
300°C rolled, alpha-annealed uranium.
Source: R. M. Mayfield, ANL.

In addition to thermal cycling growth in the alpha-phase field, cycling
through the alpha-beta transition temperature causes "wrinkling" or a surface
distortion due to different expansion characteristics of the two phases.
Internal grain boundary voids may also be produced. This happens because
there is a 1% volume expansion in going from alpha to beta uranium. The beta
phase is harder than the alpha, so the latter is plastically distorted on
heating and cooling and undergoes plastic flow, leading to the wrinkling and
voiding.

Irradiation causes growth in uranium, although the exact relationship between
thermal cycling growth and irradiation growth is not well understood.
Anisotropic growth of alpha single crystals and textured polycrystalline
material occurs at temperatures below 500°C, the growth rate increasing
steadily as the temperature falls. In single crystals of uranium, growth
occurs in the [010] direction, with an equivalent contraction in the [100]
direction and no change in the [001] direction. The net growth of a polycrystalline sample will depend on the degree of texture or preferred
orientation.

Thermal cycling and irradiation growth lead to internal stresses which, in
turn, lead to grain boundary cracking which has been a major problem in the
operation of uranium fuels, especially in the lower temperature regimes.
These internal stresses also cause the creep rate of uranium to increase
under irradiation in a manner that is linearly dependent on the neutron flux
[5].

The simplest solution to these problems is to alloy the uranium[*] so that the gamma phase is stabilized down to room temperature. However, this usually requires a considerable amount of alloying addition, e.g. 5-16% molybdenum, which absorbs sufficient neutrons to make the fuel unusable in the unenriched state in thermal reactors. The UKAEA developed "adjusted" uranium in which only small additions of alloying elements are needed to give resistance to growth. Hence, one can use natural uranium in thermal reactors, including the Magnox and French types. This is discussed in more detail in Chapter 9. On the other hand, for an application such as a fast reactor fuel, larger alloying additions are allowable and gamma-stabilized U-Mo alloys have been used (e.g. in the Dounreay Fast Reactor). The molybdenum has the additional advantage of imparting strength to the uranium.

A great deal of work has been performed on U-Mo alloys in the hope of developing a better fast reactor fuel. The body-centred cubic gamma phase can be retained by quenching U-Mo alloys of $\sim$10 wt % Mo content. Much work has been done to study the transformation of the quenched gamma phase to the alpha phase. From our point of view it is more important to note that irradiation experiments showed that dimensional stability increased with molybdenum content and was optimum at 10 wt % Mo.

Metallic uranium oxidizes fairly readily in water and in CO_2, two common reactor coolants. Thus the consequences of cladding failure become serious — the fuel oxidizes rapidly, swells progressively and locally, causing further cladding rupture. This was one reason for a move from metal fuels towards UO_2 which oxidizes much more slowly. However, one favourable property of uranium is its high thermal conductivity (at least ten times better than UO_2) which leads to small thermal gradients and low centre temperatures in the fuel. In an attempt to combine the best of both worlds, uranium intermetallic compounds were examined and U_3Si was judged to have the best combination of properties. Its resistance to water corrosion is still barely adequate and additions of aluminium have been shown to improve the corrosion resistance [6].

Adjusted uranium and its French counterpart Sicral F have performed well in gas-cooled, graphite-moderated reactors. However, such fuels are unsuitable for fast reactors; they swell too much at the high burn-ups required of a fast reactor fuel — as stated above, this led to the development of U-Mo and U-Nb "strong" fuels with less swelling. However, this line of development has been abandoned in favour of the EBR-II fuel concept in which the fuel is allowed to swell 20-30% when it releases its fission gases and swells much more slowly thereafter. A requirement of this fuel is that it should not react excessively with the cladding during normal or off-normal operation. The uranium-"fissium" fuel that is currently used in the EBR-II "driver" fuel elements has a satisfactory performance. Fissium is a synthetic mixture of fission products, designed to simulate fuel that has been recycled through the pyrometallurgical process that was developed for EBR-II. However, when one turns to the ^{238}U-^{239}Pu cycle, on which fast reactors are expected to operate, the U-Pu-fissium fuel is inadequate. This is because the temperature of reaction between the fuel and stainless steel cladding is unacceptably low — alternatively, the rate is too fast.

---

[*]An appendix on phase diagrams, that are used to describe alloys, appears at the end of this chapter.

In an effort to overcome these problems, uranium-plutonium-zirconium alloys
were developed, with U-15 wt % Pu-10 wt % Zr being the "reference" composition
[7]. This alloy has a heavy atom density of 14.4 g/cm^3 compared with
9.8 g/cm^3 for $(U,Pu)O_2$ and 12.9 for $(U,Pu)C$, hence it should give a better
breeding performance. Tests established that the alloy had adequate
compatibility with stainless steel up to 668°C over long periods of time and
to higher temperatures for shorter times. Irradiation experiments were
carried out in the EBR-II reactor [8,9] to a peak burn-up of 5.6 at %. One
element that was clad in a V-20Ti alloy continued its irradiation to 12.5 at
% burn-up without failure. An interesting feature of these irradiations was
the appearance of three concentric bands in the transverse sections of the
fuel rods (Fig. 2.7). These bands are related to the phase distributions
which existed at the temperature of irradiation. Presumably the width of
the bands will vary with fuel centre temperature and rod power. Because of
the complexity of these alloys, more research is needed before they can be
seriously considered for service in a fast breeder reactor.

Transverse Section of a Prototype Hastelloy-X
Clad U-15Pu-12Zr Alloy Fuel Element after Irradiation
to 2.4 a/o Burnup at a Maximum Cladding Temperature of 610°C (1130°F)

Fig. 2.7.

Source: ANL.

METALLIC DISPERSION FUELS

One form of dispersion fuels is two-phase metallic fuels in which the fissile phase is dispersed as small islands in a "sea" of nonfissile metal. Thus fissile effects are localized, reaction between the fuel and the coolant is essentially eliminated, and the path for heat flow from the fissile particles to the coolant is through a highly conducting medium.

The two most commonly used metallic dispersion fuels are uranium-aluminium and uranium-zirconium. Uranium-aluminium alloys are used extensively for research reactor cores where the objective is to produce a high neutron flux, hence the fuel may operate at low temperatures compared to a power reactor. Furthermore, the main limitation on neutron flux is heat transfer, because film boiling of the coolant must be avoided. Hence, research reactor fuel elements are generally of plate geometry, consisting of a "picture frame" of aluminium surrounding a twophase U-Al alloy of 15-20% uranium (Fig. 2.8). The casting and rolling of the fuel "meat" is arranged to give fuel particles (UAl_2 or UAl_3) in the size range up to $\sim$500 μm and preferably not less than $\sim$50 μm, uniformly dispersed in aluminium. Generally, such fuels are highly enriched to give a high power density, but there is a current trend towards reducing the enrichment towards 20% ^{235}U by increasing the fuel particle loading.

Like aluminium, zirconium has a low absorption cross-section for neutrons and is resistant to corrosion by water and steam. Hence U-Zr alloys have been used in water-cooled reactors, especially in the early nuclear submarine cores. As seen in Fig. 2.9, the solubility of uranium in zirconium is low, so that practical alloys consist of a dispersion of the delta phase or the $gamma_2$ phase in zirconium.

LIQUID FUELS

During the 1950s and 1960s a considerable effort was devoted to developing reactors in which the core was fluid, i.e. either a solution or a suspension which could be "reprocessed" by very direct methods and in which the problems of radiation-induced swelling, growth and fission gas build-up were eliminated. The technical problems of developing such reactors, especially the question of containment but also economics, led to their abandonment. Leading candidates were: uranyl nitrate in the ORNL Homogeneous Aqueous Reactor, uranium tetrafluoride in a eutectic molten fluoride mixture (ORNL Molten Salt Reactor), uranium-bismuth dilute solutions in the BNL/Harwell Liquid Metal Fuelled Reactor, and a uranium-plutonium-iron (or nickel or chromium) eutectic alloy in the Los Alamos Molten Plutonium Reactor Experiment. These developments are all well documented and will not be further discussed [10].

THORIUM

Thorium is a breeding element that produces ^{233}U. However, it has some attractive properties as a solvent base for uranium. It melts at 1755°C (600°C higher than uranium) and has a cubic crystal structure up to 1360°C. It is much stronger than uranium at the same temperature. All of this suggests that it should have superior radiation behaviour, which has been proven. Thorium alloys containing up to 20 wt % uranium have demonstrated superior radiation performance, e.g. a Th-20 wt % U alloy irradiated at 650°C to 4% burn-up swelled only 10% [11].

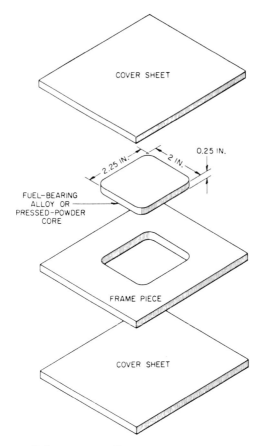

Fig. 2.8.   "Picture-frame" assembly of a research reactor
fuel element. <u>Source</u>: Uranium Oxide, by
J. Belle.

URANIUM DIOXIDE

Uranium dioxide is the most commonly used nuclear fuel today, since it is
used in all light-water moderated reactors and in most heavy-water moderated
reactors. When used as a solid solution with PuO_2 it is the reference fuel
for nearly all current fast breeder reactors. Books have been written about
its fabrication and properties [12,13], hence we will confine our attention
here to those characteristics that affect its performance in-reactor.

UO_2 is a blackish-brown refractory oxide melting at $2800^\circ C$. It has an X-ray
density of 10.97 g/cm^3 (compared to 18.5 for uranium and 13.6 for UC).
Despite this lower fissile atom density it *is* used at the natural enrichment
level as the fuel for the D_2O moderated CANDU reactors.

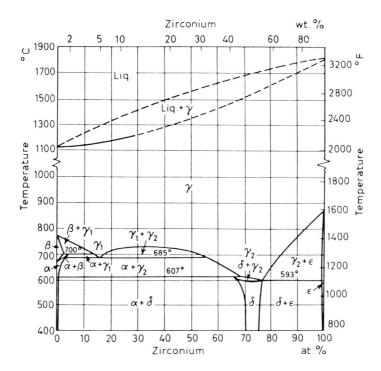

Fig. 2.9.   The uranium-zirconium phase diagram.
Source: F. A. Rough and A. A. Bauer, Battelle
Columbus Laboratories.

UO_2 has a cubic (fluorite) crystal structure and can accommodate extra oxygen
atoms in the crystal lattice at elevated temperatures. This is shown
diagrammatically in Fig. 2.10 and in Fig. A.3 of the Appendix to this
chapter. The added oxygen atoms affect its properties; of particular
importance to its performance — the creep, sintering and diffusion rates are
all increased with departure from stoichiometry. At very high temperatures
it can dissociate into UO_2 and UO_3, the latter having a high vapour pressure.
As we shall see later, the migration of UO_3 down the temperature gradient in
a "UO_2" fuel element leads to a nonrandom distribution of oxygen in the fuel
and this has an effect on cladding corrosion.

An important property of UO_2 from the standpoint of the fuel element designer
is its thermal conductivity, which is low and it varies with temperature and
under irradiation at low temperatures (Fig. 2.11). We shall discuss the
implications of this in a later chapter.

UO_2 is mainly used in the form of dense, sintered cylindrical pellets,
typically 1-2 cm in diameter. These are generally made by cold pressing and
sintering UO_2 powder (Table 2.3), although other methods such as hot pressing
are occasionally used.

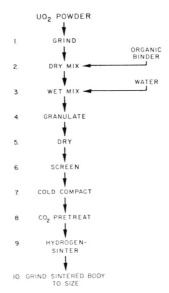

Table 2.3.   Process flow diagram of principal operations for fabrication of cold-pressed UO_2 fuel components

Source: Uranium Dioxide by J. A. Belle, USAEC/USGPO, 1961.

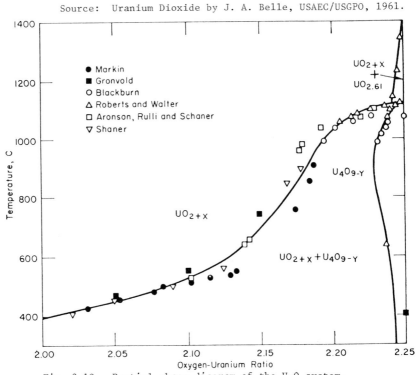

Fig. 2.10.   Partial phase diagram of the U—O system.
Source: Ceramic Fuel Elements by Robert B. Holden, pub. Gordon & Breach, NY, 1966.

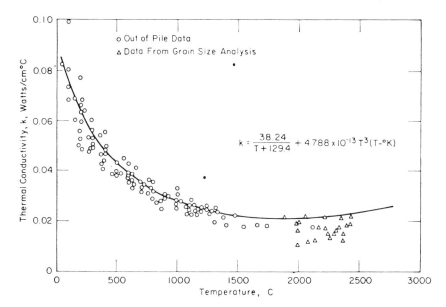

Fig. 2.11. Thermal conductivity of UO$_2$. Source: Ceramic
Fuel Elements by Robert B. Holden, pub.
Gordon & Breach, NY, 1966.

The ideal characteristics of UO$_2$ powder are:

  (i)   a small particle size,
 (ii)   regular shape,
(iii)   a "clean" surface, i.e. a minimum of adsorbed gases, and
 (iv)   a large surface area per unit mass.

To some extent (i) and (iv) are incompatible with (iii).

The characteristics of the powders obviously depend upon the nature of the
material from which they are formed. When UO$_3$ and U$_3$O$_8$ are reduced to UO$_2$
there is a large decrease in specific volume which leads to particle fracture
and an increase in surface area. Careful control of the reduction temperature
is needed to ensure that this effect is not counteracted by coalescence.
Generally, UO$_2$ powder from the reduction process is reduced in size by grinding
in a rod or ball mill for several hours. Care must be taken not to pick up
nonfissile ceramic particles in this process. This can be done by proper
choice of mill materials; rubber-lined mills are often used. In Fig. 2.12 the
effects are shown of the powder preparation route, the precipitant and the
milling procedure on UO$_2$ particle size.

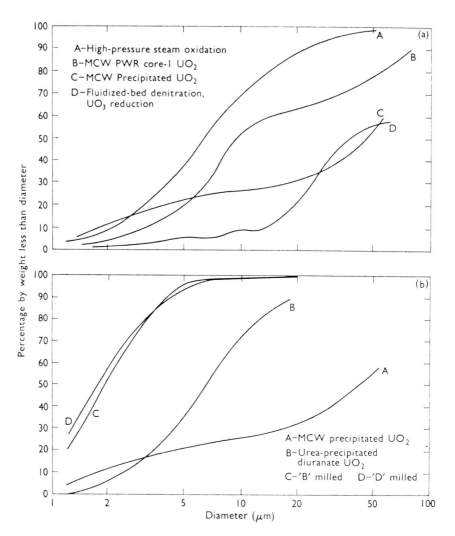

Fig. 2.12. UO_2 particle size distribution: a) the effect
of powder preparation, b) the effect of
milling procedure.   Source: B. R. T. Frost,
R. I. C. Reviews.

After milling, the powders are not generally free flowing and therefore will
not fill dies with the same quantity of material each time. The powders are
treated by the addition of 1-2% of binder, generally an organic chemical such
as polyethylene glycol, paraffin wax or polymethyl methacrylate; this may be
added dry or in solution. The powders are then granulated through sieves to
produce a free flowing powder which is poured into tungsten carbide-lined
dies and compacted under a pressure of up to 150 ton/in^2. To ensure uniform
densification of the powder and for ease of loading into stainless steel
tubes, UO_2 is generally used in the form of right cylinders. The "green"

pellets are debonded (i.e. the binder is removed) by heating in an inert or slightly oxidizing atmosphere at about 800°C. They are placed on molybdenum trays and heated in a hydrogen atmosphere in molybdenum-wound furnaces to promote sintering; several hours at 1500-1700°C are required to produce high-density pellets. Generally the aim is to produce pellets with about 5% or more of residual porosity (i.e. 95% of theoretical density or less). The porosity is then mostly within the grains and is "closed", i.e. not accessible from the outside of the pellet. The reasons for choosing this density is discussed later when properties are considered. Figure 2.13 shows the effects of various process variables on the final sinter density of UO_2.

The most common form in which ceramic fuels are used is as right cylindrical pellets inside metal tubes. However, the quest for improved performance and economics has led to the development of other fuel forms, the most important being vibrocompacted fuel and coated particles. If fuel swelling consider-ations dictate the need for a low fuel density, it may be uneconomic to make carefully sized pellets. It may be cheaper to make sintered spheres of about 500-1000 μm diameter and pack these into metal tubes; a single size of sphere gives a maximum packing density of about 74%. By adding smaller particles, ideally reducing in diameter by multiples of seven, densities up to 90% of theoretical can be made. Packing is achieved by vibrating the metal tube mechanically. Spheres may be made by the sol-gel process, which is amenable to mass production and remote control. The sol-gel route is particularly suited to ThO_2 and $(U,Th)O_2$ fuel fabrication [14].

The final form of the UO_2 fuel is a right cylindrical pellet with a density of between 85% and 95% theoretical. This material contains a number of sinter pores with diameters ranging from 10 μ down to 0.1 μ or less, which are filled with the carrier gas used in the sintering process — often nitrogen (Fig. 2.14). The size, distribution and volume of these pores exercise a strong influence over the irradiation behaviour of the fuel, as will be seen later. Hence it is important to characterize the fuel before it is tested — both in-pile and out-of-pile. Characterization should include measurements of the fuel stoichiometry, impurity content, grain size and pore size. Generally this is done on a sample from each batch of pellets [12].

$(U,Pu)O_2$

Much of what has been said for UO_2 also applies to $(U,Pu)O_2$. When used as a fast reactor fuel the composition is around 80% U and 20% Pu. Uranium and plutonium dioxides have the same fluorite crystal structure, but because of the different stabilities of valence states in U and Pu, $(U,Pu)O_2$ is generally sub- or hypostoichiometric, i.e. it is deficient in oxygen atoms. This has an influence over a number of processes that are important to irradiation behaviour, such as restructuring of the porosity, redistribution of the actinides and fuel swelling.

"Mixed" oxides, as they are generally known, may be made by coprecipitation from a U-Pu nitrate solution or may be separately precipitated from solutions and mechanically mixed, homogenization occurring during sintering and during irradiation. A limitation on mechanical mixing is imposed by Doppler coefficient safety arguments in which the size of the plutonium-rich particles must be controlled to less than ∿100 μ in order that heat transfer from Pu to U atoms be rapid during a power transient; the increasing absorption cross-section for neutrons in ^{238}U with increasing temperatures compensates for the increasing fission rate in the ^{239}Pu. An advantage of mechanical mixing is that the U and Pu streams may be kept separate until a later stage in fuel element fabrication, thereby reducing the opportunity or risk of diversion.

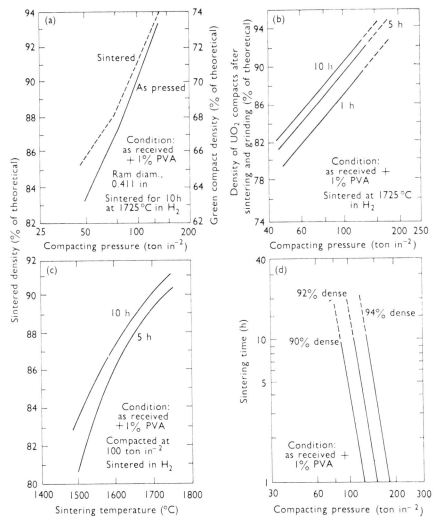

Fig. 2.13. Effects of processing variables on the
sintered density of MCW UO$_2$ compacts. (a) As
pressed and sintered density as a function of
compacting pressure. (b) Sintered density as
a function of compacting pressure and sintering time. (c) Effect of time and temperature
on sintered density. (d) Sintering time as a
function of compacting pressure and sintered
density. Source: B. R. T. Frost, R.I.C.
Reviews, 1969.

Fig. 2.14.   Electron micrograph of medium density UO_2
compact (94.4% theoretical) showing trans-
granular fracture and size and spatial
distribution of pores; 15,000X, reduction
factor, 1/2.   Source: Uranium Dioxide,
J. Belle.

CARBIDES OF URANIUM [15,16,17]

Uranium forms two carbides that are of practical interest. As the phase
diagram (Fig. 2.15 and A.4) shows, UC melts at $2780^{\circ}C$ and UC_2 at $2720^{\circ}C$.
U_2C_3 forms by a peritectoid (solid state) reaction at 2100 C. Plutonium and
thorium form similar compounds that form solid solutions with the uranium
compounds. From a practical standpoint, UC or (U,Pu)C is of interest as a
fast reactor fuel, while UC_2 or $(U,Th)C_2$ is of interest as an HTGR (High
Temperature Gas-cooled Reactor) fuel. Basically, the carbides have two
advantages over oxides: (1) their heavy metal densities are higher
(UC = 12.97 g/cm^3, UC_2 = 11.68 g/cm^3, UO_2 = 9.65 g/cm^3 of uranium);
(2) their thermal conductivities are higher by almost a factor of ten. These
lead to more favourable neutronics (breeding) and to lower thermal gradients,
respectively.

We shall defer discussion of HTGR fuels until later in this chapter, so we
will discuss the monocarbide as a fast reactor fuel at this juncture. UC has
a very narrow range of existence. A small shift towards higher uranium
content leads to the appearance of free uranium metal in the grain boundaries
(Fig. 2.16). This will lead to poor fuel performance with regard to swelling
and reaction with stainless steel cladding. A carbon-rich composition results
in the appearance of UC_2 as a second phase (Fig. 2.17) or of U_2C_3 as a second
phase. UC_2 needles appear to promote carburization of the cladding, but the
effect is much less with U_2C_3.

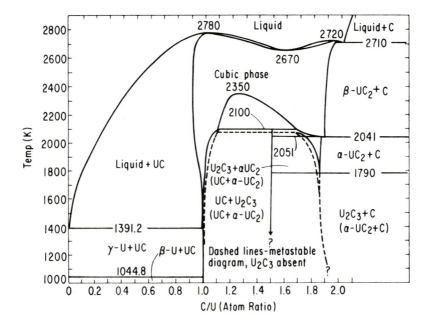

Fig. 2.15. U–C Phase Diagram. Source: ANL.

Fig. 2.16. Microstructure of hypostoichiometric U–C
            alloy, showing UC grains surrounded by uranium
            metal in the grain boundaries (Frost, *J. Nucl.
            Mat.*, 1963).

Fig. 2.17.   Microstructure of hyperstoichiometric U-C
alloy, showing UC_2 precipitates in UC grains
(Frost, *J. Nucl. Mat.*, 1963).

It must also be noted that oxygen and nitrogen dissolve readily in UC,
substituting for carbon. Hence, in arriving at the correct composition,
allowance must be made for this. One often sees the composition of carbides
defined in terms of "carbon equivalent", where the carbon, oxygen and
nitrogen contents have been added together.

UC and (U,Pu)C are made by the carbon reduction of UO_2 or $(U,Pu)O_2$. UO_2 or
$(U,Pu)O_2$ powders are mixed with the correct amount of carbon powder and
pressed into pellets. These pellets are reacted at a temperature in the
range 1400-1900°C, either in a fast pumping vacuum system (to remove the CO
product) or in an argon fluidized bed. The reacted pellets are either melted
in an arc furnace (this is now unpopular) or are ground, sieved and pressed
into pellets for final sintering, e.g. at 1500°C in pure argon. Binders and
sintering aids (especially nickel) may be used to improve the sintering
process. The typical product has an oxygen content of ∿500 ppm, a density of
95% theoretical and a grain size of ∿10-20 μm. For a fast reactor, both
(U,Pu)C fuel pellets and UC blanket pellets will be made by this route.
Carbides react rapidly with water vapour. Thus, the handling of powders and
pellets must be carried out in dry, inert atmospheres in gloveboxes.

The physical and mechanical properties of UC (and to a lesser extent (U,Pu)C)
have been measured fairly extensively and summarized [15,16,18] . (See
Table 2.4.) Routbort and Singh in ref. 18 (table 15) list the mechanical
properties that require additional study for both carbide and nitride fuels.
We will discuss these later in connection with fuel performance in a reactor.

The nitrides of uranium are very similar to the carbides as seen from the U-N
phase diagram (Fig. 2.18). The compounds melt or dissociate at different
temperatures depending on the nitrogen over-pressure. This may have
undesirable effects on reactor safety.

TABLE 2.4. Properties of Fuels

Property	U	UO_2	UC	UC_2
Melting point, $^{\circ}K$	1405	3138	2780 ± 25	2773
Density, g/cm^3	19.12	10.96	13.61	12.86
Heavy metal density, g/cm^3	19.12	9.65	12.97	11.68
Crystal structure	a	fcc (CaF_2)	fcc (NaCl)	fcc (CaF_2)
Thermal conductivity, $W/cm-^{\circ}K$	0.35 ($670^{\circ}K$)	0.03 ($1270^{\circ}K$)	0.216 (to $1270^{\circ}K$)	0.35 (to $1270^{\circ}K$)
Thermal expansion, $10^{-6}/^{\circ}K$	19 (to $920^{\circ}K$)	10.1 (to $1270^{\circ}K$)	11.6 (to $1470^{\circ}K$)	18.1 ($1970^{\circ}K$)
Electrical resistivity, ohm-cm	35×10^{-6} ($298^{\circ}K$)	1×10^3	40.3×10^{-6} ($298^{\circ}K$)	—
Specific heat, $cal/g-^{\circ}K$	0.026 (to $773^{\circ}K$)	0.065 ($700^{\circ}K$)	0.048 ($298^{\circ}K$)	0.12 ($298^{\circ}K$)
Heat of fusion, cal/mole	4760	16,000	11,700	—
Vapour pressure, atm	5×10^{-6} ($2300^{\circ}K$)	8.5×10^{-8} ($2000^{\circ}K$)	1.7×10^{-10} ($2300^{\circ}K$)	2.5×10^{-11} ($2300^{\circ}K$)
Debye temperature, $^{\circ}K$	$200^{\circ}K$	$<600^{\circ}K$, $870^{\circ}K$	—	—
Free energy of formation, kcal/mole	—	-218 ($1000^{\circ}K$)	-23.4 ($298^{\circ}K$)	—
Heat of formation, kcal/mole	—	-260 (to $1500^{\circ}K$)	-23.63 ($298^{\circ}K$)	-23 ($298^{\circ}K$)
Entropy, $cal/mole-^{\circ}K$	—	18.6 ($298^{\circ}K$)	14.15 ($298^{\circ}K$)	16.2 ($298^{\circ}K$)
Poisson ratio	0.21	0.3	0.284	—
Modulus of rupture, MPa	—	80	—	—
Modulus of elasticity, MPa	1.7×10^5	1.8×10^5	2×10^5	—
Shear modulus, MPa	0.85×10^5	0.75×10^5	0.873×10^5	—
Tensile strength, MPa	400	35	—	—
Compressive strength, MPa	—	1000	350	—
Thermal neutron fission cross-section, barns	4.18 (natural)	0.102 (natural)	0.137 (natural)	0.112 (natural)
Thermal neutron absorption cross-section, barns	7.68 (natural)	0.187 (natural)	0.252 (natural)	0.207 (natural)
Eta (η)[d]	1.34	1.34	1.34	1.34

Property	UN	Th	ThO_2	ThC
Melting point, $^{\circ}K$	3035 (1 atm N_2)	2028	3663	2898
Density, g/cm^3	14.32	11.72	10.00	10.96
Heavy metal density, g/cm^3	13.52	11.72	9.36	10.46
Crystal structure	fcc (NaCl)	fcc $1618^{\circ}K <$ bcc	Cubic (CaF_2)	Cubic (NaCl)
Thermal conductivity, $W/cm-^{\circ}K$	0.2 ($1023^{\circ}K$)	0.45 ($923^{\circ}K$)	0.03 ($1270^{\circ}K$)	0.28 (to $1270^{\circ}K$)
Thermal expansion, $10^{-6}/^{\circ}K$	9.3 (to $1270^{\circ}K$)	12.5 (to $923^{\circ}K$)	9.32 (to $1270^{\circ}K$)	7.8 (to $1270^{\circ}K$)
Electrical resistivity, ohm-cm	1.75×10^{-4} ($298^{\circ}K$)	15.7×10^{-6}	—	25×10^{-6} ($298^{\circ}K$)
Specific heat, $cal/g-^{\circ}K$	0.049 ($298^{\circ}K$)	0.038 ($970^{\circ}K$)	0.07 ($298^{\circ}K$)	0.043 ($298^{\circ}K$)
Heat of fusion, cal/mole	12,750	3300	25,000	—
Vapour pressure, atm	4.5×10^{-7} ($2000^{\circ}K$)	1.3×10^{-14} ($1500^{\circ}K$)	5×10^{-9} ($2000^{\circ}K$)	—
Debye temperature, $^{\circ}K$	—	$163.5^{\circ}K$	$200^{\circ}K$	—
Free energy of formation, kcal/mole	-64.75 ($298^{\circ}K$)	—	-279 ($298^{\circ}K$)	-6.4 ($298^{\circ}K$)
Heat of formation, kcal/mole	-70.70 ($298^{\circ}K$)	—	-293 ($298^{\circ}K$)	-7.0 ($298^{\circ}K$)
Entropy, $cal/mole-^{\circ}K$	15.0 ($298^{\circ}K$)	—	15.59 ($298^{\circ}K$)	12.0 ($298^{\circ}K$)
Poisson ratio	0.263	0.27	0.17	—
Modulus of rupture, MPa	—	—	80	—
Modulus of elasticity, MPa	—	7×10^4	14×10^4	—
Shear modulus, MPa	1.01×10^5	2.7×10^4	1×10^5	—
Tensile strength, MPa	—	230	100	—
Compressive strength, MPa	—	—	1500	450
Thermal neutron fission cross-section, barns	0.143 (natural)	—	—	—
Thermal neutron absorption cross-section, barns	0.327 (natural)	7.56	—	—
Eta (η)	—	—	—	—

[a] Orthorhombic ($<936^{\circ}K$), tetragonal ($936-1043^{\circ}K$), body-centred cubic ($>1043^{\circ}K$).

[b] Orthorhombic plus tetragonal ($<838^{\circ}K$), body-centred cubic ($>838^{\circ}K$).

[c] U containing 5% fissium (0.22% Zr + 2.5% Mo + 1.5% Ru + 0.3% Rh + 0.5% Pd). U-5% fissium is bcc above $10000^{\circ}K$, bcc + monoclinic U_2Ru between $825^{\circ}K$ and $1000^{\circ}K$, and bcc + U_2Ru + tetragonal below $825^{\circ}K$.

[d] Number of fission neutrons released per neutron absorbed.

Source: M. Simnad and J. P. Howe, *Materials for Nuclear Fission Power Reactor Technology*, Materials Science in Energy Technology (1979).

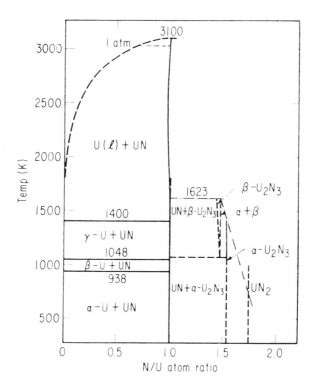

Fig. 2.18.   U-N Phase diagram.   Source: ANL.

UN has a higher heavy atom density than UC ($\sim$13.5 g/cm^3) and hence should give a better breeding performance.   This is somewhat offset by the high absorption cross-section of ^{16}N.   Separation of ^{15}N, which has a lower cross-section, would restore the breeding advantage of UN over UC, but the economics are uncertain.   Given these questions and a very similar reactor performance of UN and UC, it is concluded that UN offers no clear advantages, hence attention is directed toward UC and away from UN.

CERMETS

Cermets are dispersions of ceramic fuel particles in metal matrices [19,20]. The most common example is UO$_2$ in stainless steel.   The purpose of cermets is to overcome some of the disadvantages of UO$_2$ in order to ensure reliable fuel element performance at some economic penalty.   Hence, cermets have been studied primarily for military reactors rather than for commercial types.

A typical cermet fuel element is made by mixing UO$_2$ particles $\sim$50-500 μm diameter with stainless steel powder, sintering, placing the fuel plate in a stainless steel "picture frame" and hot rolling the assembly to final size. A typical microstructure of the fuel section is shown in Fig. 2.19.   Ideally each fuel particle should be surrounded by a web of stainless steel that is thicker than the fission product recoil distance ($\sim$10 μm), so that a continuous web of relatively undamaged steel remains.   Heat conduction is

mainly through the steel, and the temperature rise within each fuel particle
is small so that the fuel operates at all times at a relatively low
temperature. Thus fuel swelling is minimized and is strongly restrained by
the web of steel.

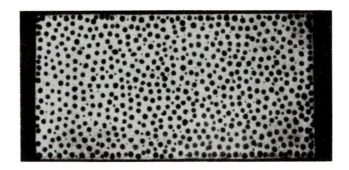

Fig. 2.19.   Cross-section of a typical UO_2-steel cermet,
10X magnification.   Source: Frost *et al.*,
3rd Geneva Conference.

In practice it is difficult to make satisfactory cermets with more than 40%
by volume of fuel.   Such a fuel has poor breeding characteristics: the steel
matrix absorbs a lot of neutrons.   Serious thought was given to using cermet
"driver" fuels for both the PFR and FFTF, but the idea was abandoned when it
was realized how poor the breeding characteristics were.   Nevertheless, a lot
was learned about how to maximize the fuel fraction and to achieve a uniform
fuel dispersion.

HTGR FUEL [21,22,23]

Another form of dispersion fuel is used in High Temperature Gas-cooled
Reactors.   In the late 1950s and the 1960s several types of dispersion fuel
were studied for use in HTGRs.   The two principal types were UO_2 in BeO
moderator and UO_2 or UC_2 in graphite.   The former was abandoned, in part
because of the poor irradiation behaviour of BeO due to helium production
and in part because of the superior performance of the graphite fuels.   The
development of the HTGR fuel is described in more detail in a later chapter.
For comparative purposes in this chapter we can consider two similar concepts
— the UK low enrichment concept and the General Atomic high enrichment thorium
cycle concept (Table 2.5).   The basic concept in each case is to coat fuel
particles with dense layers of graphite and silicon carbide to retain the
fission products and to disperse these particles uniformly throughout a
graphite moderator matrix (Fig. 2.20).   The latter may take a number of
geometric shapes (Fig. 2.21) that are variously incorporated into fuel
elements which can be stacked together to form the reactor core.

TABLE 2.5.   Comparison Between Thorium-cycle and Low
Enriched Fuel Characteristics

Characteristic	U.K. low enriched fuel	GAC thorium cycle fuel, fissile/fertile
Fuel kernel	UO_2	UC_2/ThO_2
Enrichment, %	~5	93/0
Diameter, μm	800	200/500
Porosity, %	~20	~0
Peak burn-up, % FIMA	~10	75/7
Coating	TRISO	TRISO/BISO
Thickness, μm	190	170/160
(Volume fuel) (volume coated particle)	0.27	0.05/0.22
Coated particle operating parameters		
Power per particle (average), watts	0.2	0.02/0.02
Power density (average), watts/cm^3 of particle	226	263/85
Fuel rod parameters		
Process	Overcoat/ compaction	Hot injection
Particle volume, %	<40	<65
Heavy metal loading, g/cm^3	~0.8	~0.8

The starting point for the HTGR fuel is crushed and sieved or Sol-gel particles
of UO_2 or $(U,Th)C_2$ of ~500 μm diameter. These particles are suspended in a
fluidized bed of argon and are heated inductively. Hydrocarbons and silane
are injected into the argon stream to produce different coatings on the fuel
(Fig. 2.22). First, a low-density carbon coating is produced to form a
buffer layer that absorbs swelling and provides some space for fission gases.
This is sealed by a thin, dense layer. Beyond this are two layers of dense
pyrocarbon interspersed by a layer of silicon carbide. The theoretical basis
or model for this structure will be discussed later. Suffice it to say here
that the dense pyrocarbon layers provide strength to accommodate the fission
gas pressure and the solid fission product swelling in the particle, while
the SiC layer is a diffusion barrier to the movement of the more mobile
fission products.

The technology of HTGR fuel elements has advanced to a stage where cores
containing millions of coated particles have been produced and have performed
extremely well with very low failure rates of particles.

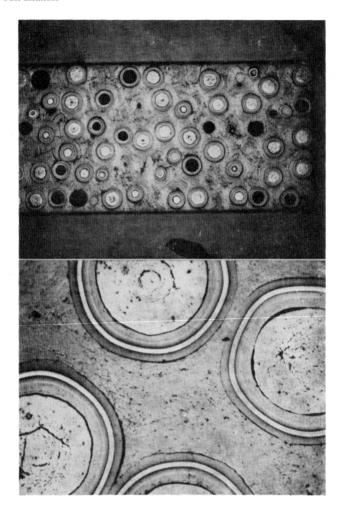

Fig. 2.20.   HTGR fuel microstructure.   Source: J. Sayers
           et al., Physical Metallurgy of Reactor Fuel
           Elements, pub. The Metals Society, 1975.

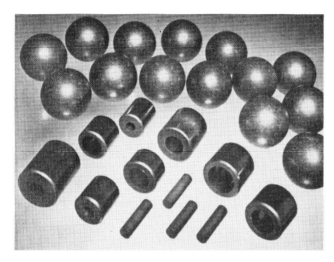

Fig. 2.21. HTGR fuel shapes. Source: M. Price and
            L. Shepherd, pub. The Metals Society, 1975.

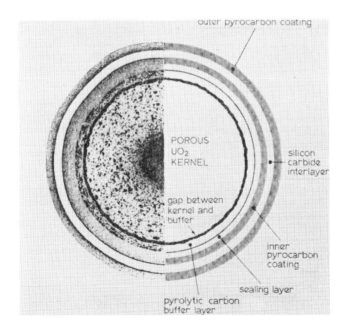

Fig. 2.22. Layers on an HTGR coated fuel particle.
            Source: M. Price and L. Shepherd, pub. The
            Metals Society, 1975.

FUEL CHARACTERIZATION

Given the need for each fuel rod or pellet to perform predictably and reproducibly, fuel characterization is a very important part of the manufacture of a fuel element. The techniques used for metal fuels, dispersion fuels and ceramic fuels differ in some respects. Metal fuels are examined by nondestructive methods, such as radiography, for casting defects and by X-ray diffraction methods for texture. Optical metallography is used extensively to measure grain size, inclusions and microscopic defects. This is often supplemented by electron microscopy using both replica and thin-film techniques, although the latter is of more use in studying irradiated samples.

Ceramic fuel powders are characterized by measurements of surface area and absorbed gases or surface chemistry, since these strongly influence the nature of the sintered product. Pellets are examined by a porosimeter to determine the amount of open porosity (by difference). Selected pellets are destructively examined for stoichiometry, although this can also be done nondestructively by X-ray diffraction. Optical and electron microscopy are used to determine the grain size and pore size distributions.

In all cases wet chemical analyses are performed on representative samples from each production batch, and nuclear assay methods, both destructive and non-destructive, are used to measure fissile atom concentrations and isotopic ratios.

In all cases "archive" samples must be retained so that valid comparisons of irradiated and unirradiated fuel may be made at the end of irradiation experiments or of power reactor core lifetimes.

REFERENCES

1. Elliott, David M. and Weaver, Lynn E. (eds.), *Education and Research in the Nuclear Fuel Cycle*, University of Oklahoma Press (1972).
2. Gittus, J. H., *Uranium*, Butterworths (1963).
3. Harrington, C.D. and Ruehle, A. E., *Uranium Production Technology*, Van Nostrand, New York (1958).
4. Mayfield, R. M., *Trans. ASM*, 50, 926 (1958).
5. Cottrell, A. H., *Metals Rev.* 29, 479 (1956).
6. Domagala, R. F. *et al.*, paper to International Meeting on Reduced Enrichment Fuel, Argonne (November 1980).
7. Walter, C. M., Golden, G. H. and Olson, N. J., Argonne National Laboratory Report ANL-76-28 (1975).
8. Kittel, J. H. *et al.*, *Nucl. Eng. Design*, 15 (4), 373 (May 1971).
9. Murphy, W. F. *et al.*, Argonne National Laboratory Report ANL-7602 (1969).
10. Lane, J. A., MacPherson, H. G. and Maslan, Frank, *Fluid Fuel Reactors*, Addison-Wesley, Reading, MA (1958).
11. Kittel, J. H. *et al.*, Argonne National Laboratory Report ANL-5674 (April 1963).
12. Belle, J. (ed.), *Uranium Dioxide: Properties and Applications*, US Government Printing Office (1961).
13. Holden, R. B., *Ceramic Fuel Elements*, Gordon & Breach (1966).
14. Clinton, S. D. *et al.*, Oak Ridge National Laboratory Report ORNL-2965 (1961).
15. Frost, B. R. T., *J. Nucl. Mat.* 10, 265 (1963).
16. Accary, A., *J. Nucl. Mat.* 8, 281 (1963).
17. Leary, J. and Kittel, J. H. (eds.), *Proceedings of a Topical Meeting on Advanced LMFBR Fuels*, American Nuclear Society (1977).

18. Routbort, J. L. and Singh, R. N., *J. Nucl. Mat.* <u>58</u>, 78 (1975).
19. Frost, B. R. T. *et al.*, *Proceedings of the 3rd Conference on Peaceful Uses of Atomic Energy*, Geneva (1964).
20. Weber, C., *Progress in Nuclear Energy*, Series 5, Vol. 2 (1959).
21. Gulden, T. D., Harmon, D. P. and Stansfield, O. M., *Proceedings of the International Conference on Physical Metallurgy of Fuel Elements*, The Metals Society, London, pp.410–415 (1975).
22. Price, M. S. T. and Shepherd, L. R., ibid, pp.397–409.
23. Sayers, J. B., Knowles, A. N., Horner, P. and Sawbridge, P. T., ibid, pp.423–430.

APPENDIX.   PHASE DIAGRAMS

Phase diagrams (also known as constitutional diagrams or equilibrium diagrams) are graphic representations of the phases that are formed between two or more elements over a wide range of temperatures. They are based on the Phase Rule which states the number of phases that can exist in equilibrium with one another. A phase is a clearly defined crystalline form or a liquid form. The phase rule of Gibbs is written as:

$$f = c - p + 2$$

where f = the degrees of freedom or variance, which may include temperature, pressure and concentration,
  c = the components (in this case the number of elements),
  p = the phases, e.g. different crystal structures.

For the case considered most commonly f = 2 (solid and liquid) and c = 2 elements. Then the phase rule becomes:

$$f = 2 - p + 2,$$
$$= 4 - p.$$

Thus there is a single phase where the system is trivariant — the temperature, pressure and composition must be exactly specified. When composition is allowed to vary (is not fixed), two phases may exist. For practical purposes we are concerned mainly with two- and three-components systems in which concentration and temperature are the variables.

Phase diagrams can be extremely useful in fuel element technology as long as it is remembered that they represent equilibrium states, i.e. the result of reactions taken to completion. In practice we are not often dealing with equilibrium conditions, so allowances must be made.

A complication in uranium and plutonium phase diagrams is the fact that these elements exist in several distinct crystalline structures in the solid state. Uranium has three allotropes, designated α, β and γ, while plutonium has six allotropes (see Table A.1). This complicates phase diagrams as seen in Fig. A.1 for the uranium-aluminium system. Uranium and aluminium form three intermetallic compounds, δ = UAl_2, ε = UAl_3 and ζ = UAl_4. The diagram tells us that these melt at 1590°, 1350° and $730^\circ C$, respectively. UAl_2 or δ forms solid mixtures with α-U up to $655^\circ C$, with β-U from 655° to $750^\circ C$ and with γ-U from 750° to $1105^\circ C$. $1105^\circ C$ is not the melting point of uranium — it is $1133^\circ C$, but a small amount of aluminium depresses this melting point down to a eutectic, or minimum melting point mixture, at $1105^\circ C$.

TABLE A.1. Crystal Structures and Transformation Temperatures
of the Plutonium Allotropes[a]

Allotrope	Temperature range of stability, °C	Space lattice and space group	Unit cell dimensions, Å	Atoms per unit cell	X-ray density, g/cm³
α	below ~ 115	primitive monoclinic $P2_1/m$	(21°C) $a = 6.183$ $b = 4.822$ $c = 10.963$ $\beta = 101.79°$	16	19.86
β	~ 115 - ~ 200	body-centred monoclinic $I2/m$	(190°C) $a = 9.284$ $b = 10.463$ $c = 7.859$ $\beta = 92.13°$	34	17.70
γ	~ 200 - 310	face-centred orthorhombic Fddd	(235°C) $a = 3.159$ $b = 5.768$ $c = 10.162$	8	17.14
δ	310 - 458	face-centred cubic Fm3m	(320°C) $a = 4.6371$	4	15.92
δ'	458 - 480	body-centred tetragonal I4/mmm	(465°C) $a = 3.34$ $c = 4.44$	2	16.00
ϵ	480 - 641	body-centred cubic Im3m	(490°C) $a = 3.6361$	2	16.51

[a]From W. H. Zachariasen and F. H. Ellinger, *Acta Cryst.* 16, 780 (1963); ibid.,
p. 369; W. H. Zachariasen and F. H. Ellinger, *Acta Cryst.* 8, 1431 (1955); and
F. H. Ellinger, *Trans Met. Soc. AIME*, 206, 1256 (1956).

URANIUM-ALUMINUM

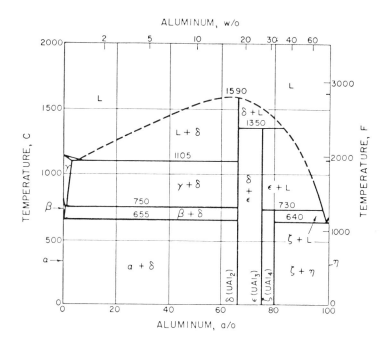

Fig. A.1.

The uppermost curve defines the temperatures and compositions at which U-Al alloys become wholly liquid and is called the "liquidus". Below it solid begins to separate out until at 1105°C alloys containing up to 66 a/o Al are completely solid, thus the 1105°C horizontal line is the "solidus". Note in the U-Al diagram that the abscissa is plotted in atomic percent while a scale for weight percent is plotted along the top.

Another interesting diagram is uranium-iron (Fig. A.2), because it can tell us what may happen when uranium metal is heated in contact with iron (or steel). The melting point of uranium is depressed to a eutectic at 725°C, which is not much above some reactor coolant outlet temperatures. Note that the eutectic solidifies to form two U-Fe compounds: U_6Fe and UFe_2.

In the two examples given so far, the compounds form at very specific atomic ratios. In some cases the compound can exist over a range of composition. This is true for the important compound UO_2. Figure A.3 shows the phase diagram plotted in terms of oxygen : uranium ratio (O/U). Above 300°C UO_2 can exist to higher O/U ratios: up to about 2.15 at 900°C. Note that the other compounds U_4O_9 and U_3O_8 can also exist over ranges of O/U. Interestingly, UO_2 can accommodate additional oxygen atoms (but not less than O/U = 2.0) while U_3O_8 displays the reverse effect. These facts have profound effects on the properties and behaviour of these compounds. Since UO_2 is processed at high temperatures and is rapidly cooled, it can retain the UO_{2+x} structure at room temperature. $UO_{2.0}$ is the "stoichiometric" composition. $UO_{2.1}$ is nonstoichiometric, often referred to as hyperstoichiometric (above the stoichiometric composition), while U_3O_{8-x} is hypostoichiometric. PuO_2 behaves like U_3O_8, i.e. it displays hypostoichiometry and it tends to dominate in UO_2-PuO_2 mixtures which are hypostoichiometric.

URANIUM-IRON

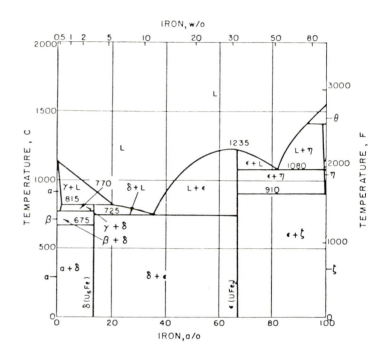

Fig. A.2.

Uranium-carbon, on the other hand, displays no departures from stoichiometry (Fig. A.4). Three compounds are formed: UC, U_2C_3 and UC_2. U_2C_3 dissociates at $\sim$1800°C in a complex manner.

Uranium and plutonium metals form a complex phase diagram (Fig. A.5) because of their many phases. Note that the two body-centred cubic phases ε-Pu and γ-U form a continuous solution, while the dissimilar structures cannot dissolve in each other.

While binary (two-element) diagrams are most common, ternary and higher phase diagrams have been drawn up. However, graphics become a problem because we need three-dimensional diagrams to represent ternary systems and this is difficult on paper. Two expedients are used. Most commonly, a diagram is drawn for each of a series of temperatures. Thus Fig. A.6 shows ternary "sections" for the U-Nb-Zr phase diagram at 500°C and at 630°C. The lines define the boundaries of the various solid phases that exist at this temperature.

URANIUM-OXYGEN

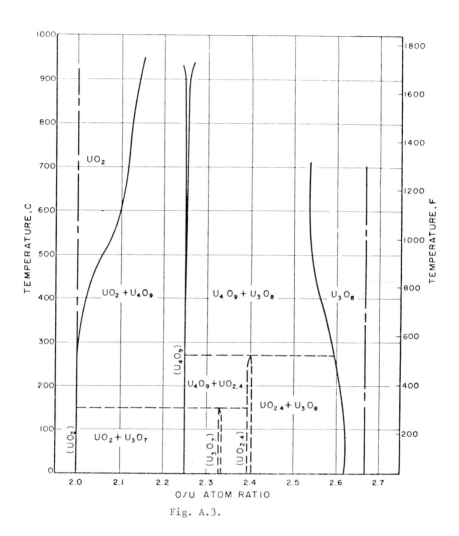

Fig. A.3.

URANIUM-CARBON

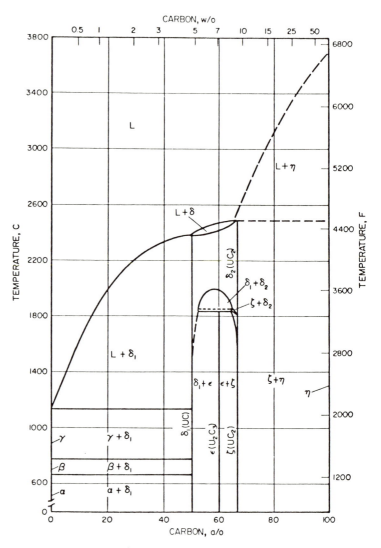

Fig. A.4.

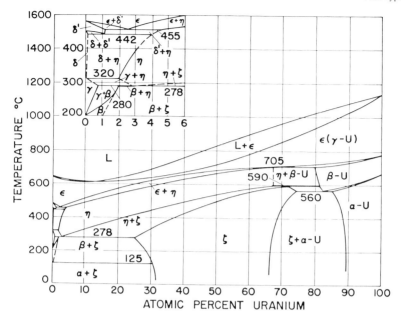

Fig. A.5.

Alternatively one can select a particular combination of two elements, e.g. U-34 a/o Ti in Fig. A.7, and make a vertical "cut" of the diagram to produce a pseudobinary system in which one can plot the composition-temperature relations. One may construct a number of these to define the full diagram.

Two publications of actinide element phase diagrams are of particular use:

1. *Constitution of Uranium and Thorium Alloys*, by Frank Rough and Arthur Bauer, USAEC Report BMI-1300 (1958).
2. *Constitution of Plutonium Alloys*, by F. H. Ellinger, W. N. Miner, D. R. O'Boyle and F. W. Schonfeld, USAEC Report LA-3870 (1968).

URANIUM-NIOBIUM-ZIRCONIUM

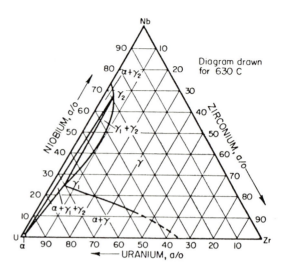

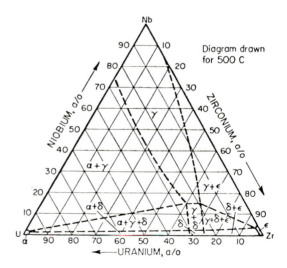

Fig. A.6.

URANIUM-MOLYBDENUM-TITANIUM

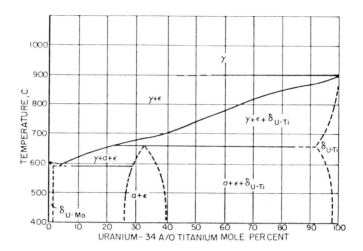

Fig. A.7.

CHAPTER 3

Irradiation Behaviour of Fuels

In this chapter we will consider the phenomena that occur in a nuclear fuel during and after fission. It is obvious that a nuclear fuel is irradiated in a fuel element in a reactor core and that the fuel element geometry and heat transfer characteristics will strongly influence the performance and behaviour of the fuel. In this and the following chapter it is intended to lay the groundwork for a discussion of the overall behaviour of fuel elements in Chapter 5.

Since this is not a book on reactor physics, the reader is merely reminded that it is the odd-numbered isotopes of uranium and plutonium that are fissionable — principally ^{235}U and ^{239}Pu. The fission cross-sections decrease as the neutron energy increases (Fig. 3.1). Thus, a thermal (moderated) reactor requires a lower concentration of fissionable isotopes than a fast (unmoderated) reactor [1]. However, the basic nuclear reaction is similar in the two cases: absorption of a neutron renders the nucleus unstable and it fissions, producing two fission fragments, neutrons, gamma rays and excess energy (Fig. 3.2). The energy results from the mass imbalance in the fission reaction and equals ∿200 MeV per fission. This energy, which is distributed between the two fission fragments, together with lesser energy in the neutrons and gamma rays, constitutes the nuclear energy which is captured and converted to heat and electricity. Let us first consider how this energy is transferred to the crystal lattice of the fuel. Fission fragments and neutrons move rapidly through the lattice, exchanging their energy with the lattice atoms until they come to rest [2,3]. The energy exchange mechanisms are represented schematically in Fig. 3.3 for neutrons. The energy carried by the fission fragments predominates and within the fuel we can probably neglect the neutrons, but the damage mechanisms are similar. Each fragment is highly ionized (about 20+) and has a range of about 7–10 μm in fuel materials. Over this range it excites atoms up to 100 Å radius from the track centre, equivalent to raising their temperature to thousands of degrees locally; this is known as a "fission spike" [4]. These are equivalent to the thermal spikes in Fig. 3.3. Within each spike a number of atoms are displaced from their lattice positions to produce single vacancies and interstitials. As their names imply, a *vacancy* is a vacant lattice site from which an atom is ejected into an *interstitial* site between the regular lattice sites. These are the two basic types of lattice defect. If the temperature is high, these will recombine. If the temperature is low, they remain as

50

single defects and have a large effect on transport processes such as
thermal and electrical conductivities. At intermediate temperatures they may
cluster and collapse into loops. Such effects are observable in a number of
ways, e.g. by X-ray lattice parameter measurements, by transmission electron
microscopy and by electrical resistivity measurements. Defects have practical
importance to fuel element behaviour in relation to their effect on thermal
conductivity and they are significant in relation to gaseous fission product
behaviour and to diffusion processes. In the latter context, diffusion is
controlled by vacancies and fission produces an abundant supply of these.
It is possible, therefore, that diffusion-controlled processes such as creep
and sintering may be enhanced while fission is proceeding.

Once the fission fragments have come to rest, they are more usually described
as fission products (fp). Their abundance varies with atomic weight as shown
in Fig. 1.7. Furthermore, they are radioactive and undergo decay processes
(Fig. 3.2) and they may capture neutrons so that the prediction of their
concentration at some arbitrary point during irradiation is difficult. There
are some significant differences between the yields from uranium and
plutonium fission: these are visible in Fig. 1.8.

The abundant fission products may be grouped conveniently according to their
physical and chemical characteristics:

Noble metals: Mo, Tc, Rh, Ru, Pd
                                                (At high oxygen potentials Mo may
                                                exist as MoO_2 or MoO_3)

Readily oxidized but insoluble in UO_2:   BaO, SrO
Readily oxidized but soluble in UO_2:   Zr, Ce, Nd
Volatiles: Cs, I, Br, Te
                                                (Some of these may be present as
                                                cesium halides)

Stable gases: Xe, Kr

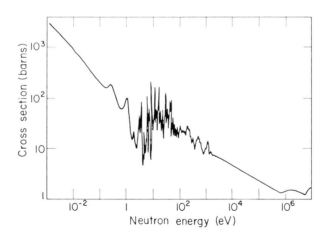

Fig. 3.1. The overall dependence of ^{235}U fission cross-
           section on neutron energy is moderately smooth,
           except for the region where resonances cause
           the cross-section to vary rapidly with energy.
           Courtesy: Lawrence Berkeley Laboratory.

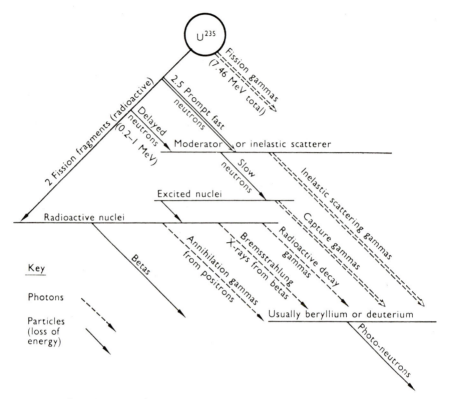

Fig. 3.2. Various types of radiation found in a nuclear
reactor. <u>Source</u>: B. R. T. Frost, R.I.C.
Reviews.

The detailed study of fission products is important for several reasons: (a)
they must be removed in reprocessing and stored or used as radioactive tracers
or heat sources, (b) they cause the fuel to swell, (c) the gases, if released
from the fuel, create a pressure, (d) some may react with cladding materials,
and (e) they represent the source term in reactor siting and licensing
arguments. Swelling and gas release are the most important effects in
determining the behaviour of fuel elements.

The "basic" swelling of a fuel is the net volume increase incurred when all
of the fission products are atomically dispersed but are able to attain their
equilibrium oxidation states. In UO_2, for example, the solution of zirconium
and the rare earth elements in the lattice causes a volume decrease due to
atomic size and charge effects, while caesium, molybdenum, the noble metals
and the insoluble oxides (see above) cause volume increases. Attempts have
been made to calculate from first principles these net swelling rates.
Davies [5] and Wait [6] arrived at a volume of 0.5% volume increase per 1%
burn-up (of heavy atoms) while Anselin and Baily [7] derived a value of 0.3%.
Davies also calculated a value of 0.9% for UC. These calculations are very
sensitive to the chemistry of the fuel. For example, the chemical state of
the fission products will depend on the oxygen potential in the fuel, which
in turn depends on the initial ratio of oxygen to uranium atoms (O : M ratio)
in UO_{2+x}.

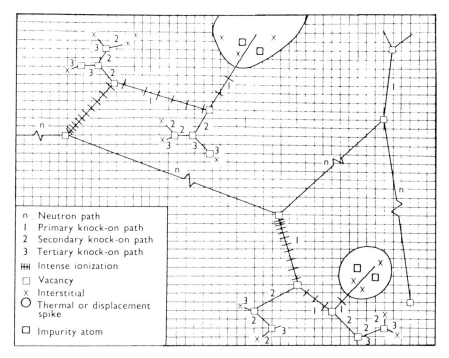

Fig. 3.3. Five mechanisms of radiation effects.
Represented are intense ionization, vacancies,
interstitials, impurity atoms and thermal or
displacement spikes. Grid-line intersections
are equilibrium positions for atoms. Source:
B. R. T. Frost, R.I.C. Reviews, 1969.

Conversely, one is interested in the oxygen potential or O : M ratio as a
function of fuel burn-up, because this may influence possible chemical
reactions with the cladding. Calculations have been performed to predict
the change of O : M with burn-up for uranium and for uranium-plutonium fuels
[8]. In general terms the O : M ratio increases with burn-up (Fig. 3.4),
exacerbating reactions with cladding materials. However, as we will discuss
later, the existence of a thermal gradient across the fuel gives rise to
oxygen migration so that an irradiated fuel pellet will show a variation of
O : M ratio from its centre to its surface [8]. (Figure 3.5 plots calculated
variations for $(U,Pu)O_2$ of varying initial O : M ratio.) It is possible to
measure this variation by analysis of small particles of molybdenum and/or
MoO_2 distributed through the fuel. The amount of MoO_2 indicates the local
oxygen potential. We will return to this topic later when discussing fuel-
cladding compatibility.

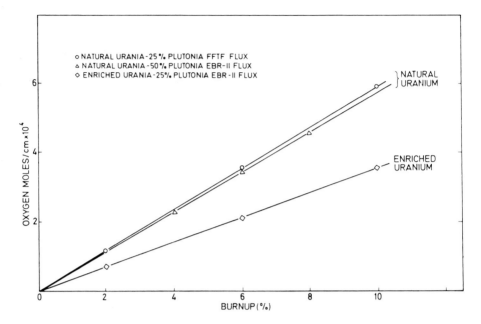

Fig. 3.4.   The production of excess oxygen due to fission
of UO_2 with a natural urania-plutonium solution
and an enriched urania-plutonia solution.
Source: C. Johnson, ANL.

The gaseous fission products are very insoluble in all fuels and hence they
have a strong tendency to agglomerate to form bubbles. Generally, this
happens fairly rapidly, so that the performance of fuels is very dependent on
the bubble behaviour. Hence, the physics of bubble behaviour in fuels has
been extensively studied. Basically bubbles will either grow and contribute
significantly to fuel swelling or they will migrate to free surfaces where
the gas is released into the volume bounded by the cladding. The extent to
which bubbles migrate depends on the severity of the thermal gradient from
the fuel surface to its centre, which depends on the rate of fission, but
particularly on the thermal conductivity of the fuel.

Metallic uranium and uranium alloys have a high thermal conductivity and
therefore generate relatively small thermal gradients, typically $\sim 100^{\circ}$C/cm.
UO_2, on the other hand, is a very poor conductor and generates large thermal
gradients — typically 2000°C/cm. Carbides are an intermediate case, with
gradients typically around 500°C/cm. These facts have a major effect on
bubble migration and thus on fuel behaviour.

From a fuel element design standpoint we can view fission gas behaviour in
two different ways: (1) we can measure the swelling and gas release in fuel
elements over a range of parameters and develop empirical relationships, or
(2) we can develop a physical understanding of the phenomenon and apply this
understanding to design problems. In practice, as one would expect, we do
both.

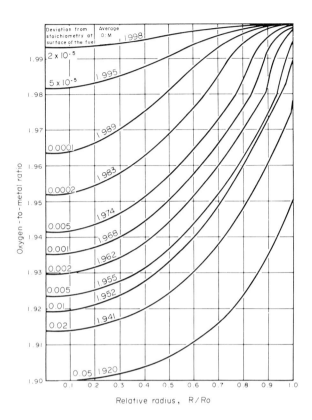

Fig. 3.5. Oxygen : metal ratio profiles across $(U,Pu)O_{2-x}$
          pellets as a function of initial O/M ratio.
          Source: E. Aitken and S. K. Evans, IAEA, 1974.

The radiation swelling of uranium metal [9] is plotted in Fig. 3.6 as a
function of burn-up. One can see the calculated contribution of the solid
fission products (dotted line). Of the remaining swelling it is believed
that 80% is due to fission gas bubbles ∿0.1 µm in diameter spaced ∿1 µm
apart and 20% is due to grain boundary cracks caused by a combination of
thermal cycling and radiation embrittlement. Under certain circumstances
"breakaway swelling" may occur where the combination of thermal cycle
cracking and fission gas accumulation can cause excessive swelling [10].

If one irradiates metallic fuels to very high burn-ups one gets a gas
release-swelling relationship of the type shown in Fig. 3.7 [11]. That is,
when the fuel swelling exceeds 20%, the fission gas release becomes very high.
The reason for this is that at 20% swelling there are sufficient bubbles on
the grain boundaries to connect up and provide a continuous path for the
escape of fission gases, after which swelling is drastically reduced. This
discovery formed the basis for the design of the EBR-II uranium-fissium
"driver" fuel. The fuel is separated from the cladding by a sodium bond
which ensures a good heat transfer. The bond is large enough to allow the
fuel to swell by more than 20% before it touches the cladding, by which time
the swelling rate has reduced to a low value. The sodium is displaced into
a plenum which must also accommodate the released fission gases.

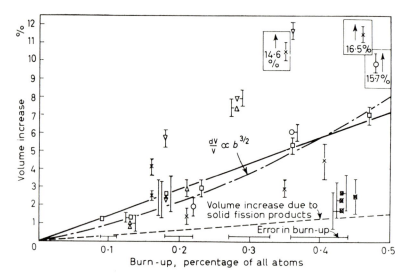

Fig. 3.6.   Swelling of uranium as a function of burn-up.
             <u>Source</u>: Uranium by J. H. Gittus, Butterworths,
             1963.

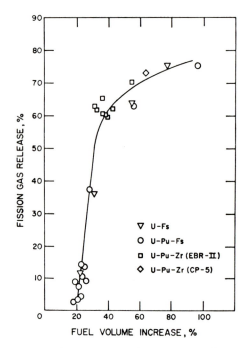

Fig. 3.7.   Swelling of metallic fuels at high burn-ups
             versus fission gas release.   <u>Source</u>:
             W. N. Beck *et al.*, ANL.

At one time it was thought that the swelling of metal fuels was sensitive to the fuel strength. Hence it was thought that alloying elements like molybdenum reduced swelling by providing strength. However, it became recognized that the main restraining force on gas bubbles was surface tension rather than the matrix strength. Hence there is more incentive to keep bubbles small, e.g. by providing numerous nuclei on which they can form, rather than strengthening the matrix. Of course, molybdenum fills another role in stabilizing the cubic beta phase which is not susceptible to thermal cycling growth.

In a fuel element in a reactor core, heat flows from the centre of the fuel rod or pellet towards the coolant, i.e. there is a radial thermal gradient. Bubbles will migrate up the thermal gradient and will grow larger as they move towards the centre. Bubble movement will be checked by various obstacles in the fuel, especially dislocations, precipitates and grain boundaries. If bubbles are hung up on grain boundaries for long enough times they may touch and form a pathway to the outside of the fuel, permitting the release of the fission gases from the fuel. While this is going on fission fragments are bombarding the bubbles and blasting them back into the fuel lattice ("resolution"). Hence, there is a dynamic equilibrium between bubble growth and migration on the one hand and resolution on the other. Models of these processes have been developed and will be discussed later.

Metal fuels operate at low fractions of their melting points where fission gas diffusion is fairly slow. Furthermore, metal fuels have high thermal conductivities so that thermal gradients are small and the driving force for directed bubble migration is small. Bubbles generally remain quite small and gas release is negligible. However, experiments on pure uranium showed that breakaway swelling of the order of 100% could occur. This was due in part to the grain boundary cracking discussed above, but possibly also to bubbles touching and coalescing suddenly. It was discovered that the addition of a few hundred parts per million of iron and aluminium combined with a heat treatment (quenching from the beta phase) removed breakaway swelling. This material is known as "adjusted uranium" and is the standard fuel in the UK Magnox reactors. The mechanism of stabilization is believed to be the effect of fine precipitates preventing bubble and grain boundary movements.

OXIDES

Because oxides are the standard fuel for LWRs and for fast breeder reactors, and because their irradiation behaviour is complex, they will be discussed at some length.

The complications arise from the low thermal conductivity of UO_2 and $(U,Pu)O_2$ which gives rise to very large radial thermal gradients in the fuel pellets, often of the order of $2000-4000^{\circ}C/cm$. This gradient causes the fuel microstructure to change quite rapidly after the reactor power has been raised to its operating level. That is to say, the original microstructure of a sintered oxide pellet (consisting of equal sized grains containing pores) changes fairly rapidly with time. Figure 3.8 shows an electron microscope replica of unirradiated UO_2. The sinter pores appear as "tails", because they are revealed by a shadowing process. After some time at full power the cross-section of a fuel element has the appearance shown in Fig. 3.9(a). Parenthetically it should be noted that this figure represents the state of the fuel *after* it has cooled down and has been cut and polished in a hot cell. The cross-section consists of several "zones" as shown schematically in Fig. 3.9(b). These zones have fairly well-defined temperatures associated with their boundaries; these are noted in the figure.

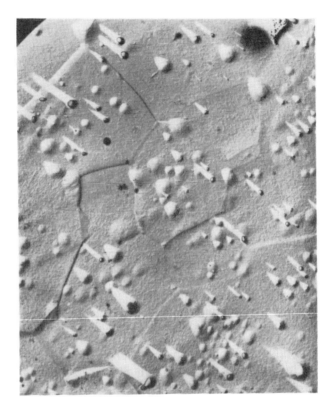

Fig. 3.8.   Surface replica electron micrograph of
polished and acid-etched UO_2 showing closed
porosity in grain interiors and general grain
structure, 4000X.   Source: B. R. T. Frost and
B. T. Bradbury.

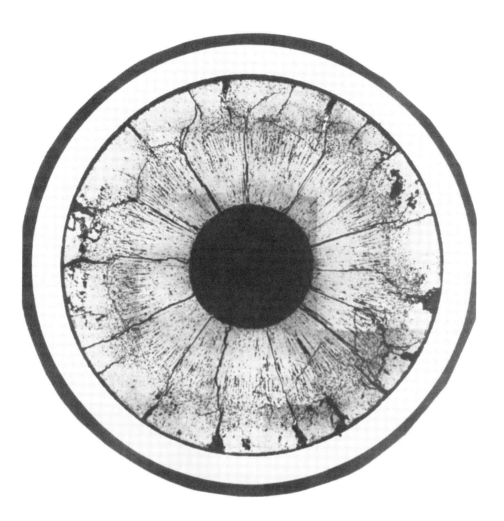

Fig. 3.9(a) Cross-section of an irradiated UO$_2$ fuel
element, showing restructuring. ANL.

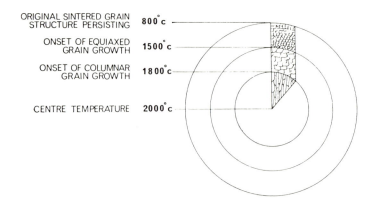

ORIGINAL SINTERED GRAIN
STRUCTURE PERSISTING   800°c

ONSET OF EQUIAXED
GRAIN GROWTH   1500°c

ONSET OF COLUMNAR
GRAIN GROWTH   1800°c

CENTRE TEMPERATURE   2000°c

Fig. 3.9(b)   Schematic representation of macrostructure
developed in a highly rated UO_2 pellet.
Source: B. R. T. Frost.

The fuel restructuring involves two types of grain growth. Above ∿1500°C the grains grow uniformly to produce larger equiaxed grains. The growing grains sweep up the sinter pores and trap them in the boundaries. Above ∿1800°C directional grain growth begins, leading to the formation of long and narrow columnar grains that are oriented towards the hot centre of the fuel. These are believed to form by two possible mechanisms: either by the migration of pores up the temperature gradient or by a solid state diffusion process. In the former process it is envisioned that UO_2 on the hotter side of a large pore evaporates and condenses on the cooled side, creating a transport process for the pore up the thermal gradient. The movement of the pores to the centre of the fuel results in densification of the grains and the formation of a central hole or "void". The grain densification causes the fuel thermal conductivity to rise a little and to cause the centre temperature to drop a little.

It can take as little as 24 hours at power for the restructuring to occur. Thereafter more gradual changes occur, which are more related to the irradiation processes. However, a purely mechanical effect does occur; when reactor power is changed rapidly the temperatures in the fuel change rapidly, leading to thermal stress changes and to fuel cracking. Such cracks are clearly visible in Fig. 3.9(a) — they probably formed at the final shutdown before the fuel was removed from the reactor. During prolonged operation at power, cracks will heal due to diffusion processes that are enhanced by fission. Hence, the fuel microstructure changes continuously with time (and burn-up). For a more detailed and rigorous discussion of the restructuring processes the reader is referred to a paper by Nichols [12].

A structural change that occurs early in life is densification. This effect occurs in both thermal and fast reactors, but has much greater technological significance in thermal reactors. When UO_2 fuel elements were unloaded from some power reactors in 1972 the cladding was seen to have collapsed due to coolant pressure in regions where fuel pellets should have been. On examination it was found that the pellet stack had shrunk by several inches over a 12-ft length [13]. Figure 3.10 gives the results of some actual stack

length changes with burn-up. More detailed examination showed that the
individual pellets had shrunk or densified under irradiation during the early
stages of operation at power in the reactor. Electron microscope studies of
the densified pellets showed that the smallest pores (<1 μ) had disappeared
from the regions that had not experienced grain growth* and that there was
a good correlation between densification and the initial pore fraction in
this size range. The problem was addressed in two ways: first, fuel pins
were prepressurized with helium to a level equal to or greater than the
coolant pressure to prevent cladding collapse; this has worked well. Second,
the fuel manufacturers altered their fabrication methods to eliminate the
fine pores from the finished microstructure — early evidence shows that this
also works well.

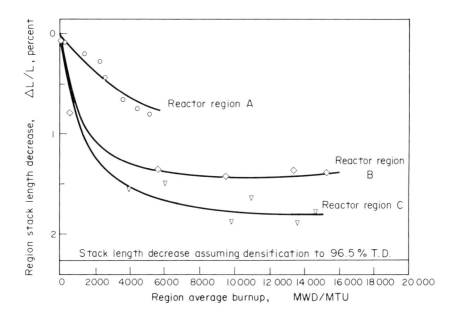

Fig. 3.10. Neutron detector measurement of stack length
changes for 92% TD UO-2. Source: R. O. Meyer,
NUREG-0085.

The problem has been examined from a mechanistic standpoint and it seems very
likely (but difficult to prove) that the small pores are destroyed by a
combination of fission tracks passing through or near to these pores and the
rapid migration of interstitial atoms to those pores. The NRC has developed
a densification model, described in ref. 13, which is based on the fact that
resintering of production UO_2 pellets at 1700°C for 24 hours gives a very
similar densification to that found in-reactor. Figure 3.11 shows the degree
of correlation between the two processes. The NRC model states that between
20 and 2000 MWD/tU the fuel density change

$$\Delta\rho = m \log (\text{burn-up}) + b \qquad (3.1)$$

---

*LWR fuel centre temperatures are generally below 1500°C.

where m and b are given by:

$$0 = m \log (20) + b \tag{3.2}$$

and

$$\Delta \rho_{max} = m \log (2000) + b. \tag{3.3}$$

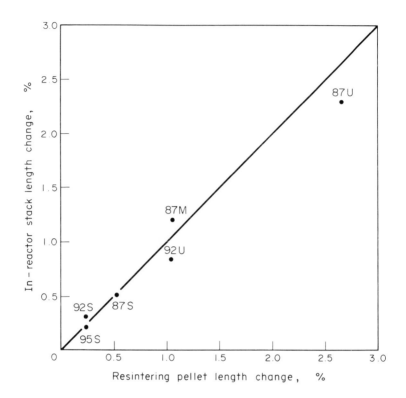

Fig. 3.11. In-reactor stack length changes compared with
out-of-reactor resintering length changes
adjusted to correspond to resintering at
1700°C for 24 hours. Source: R. O. Meyer,
NUREG-0085.

In Fig. 3.12 the solid lines are derived from these expressions and are
compared with experimental results from tests in the Halden reactor [15].

Let us now consider fission gas and the other fission product behaviour in
oxide fuels after the initial restructuring and densification phase. It is
important to remember that the early interest in UO_2 was in connection with
thermal reactor fuels, especially the CANDU and gas-cooled systems (AGR and
its American counterpart the EGCR), in which plenums were not a viable
option and hence low gas release was a goal. The emphasis on fission gas
release studies was on behaviour at fuel temperatures below that of grain
growth. Three types of measurement were common:

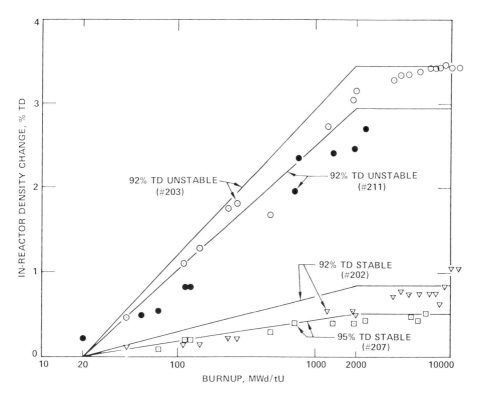

Fig. 3.12.   In-reactor densification of high density
Halden fuels as a function of burn-up. Solid
lines correspond to the kinetics model with
the maximum density change estimated from the
data shown. <u>Source</u>: R. O. Meyer, NUREG-0085.

1. Continuous measurement in-pile, often called the "sweep-capsule" method,
   in which helium is passed over a heated fuel sample and then out of the
   reactor. Generally, the helium deposits its entrained fission products
   in a cold charcoal trap which is counted for the rate of accumulation of
   the different isotopic species.
2. Puncture of fuel pins after irradiation followed by measurement of the
   volume of the released gas.
3. Post-irradiation annealing studies similar in concept to the in-pile
   swept capsule method.

For the more highly rated, mixed oxide fuels being developed for fast
reactors similar tests have been used, although post-irradiation annealing
has fallen out of favour.

For thermal reactor fuels it was thought that two basic mechanisms of gas
release operated below 1600°C; up to 1000°C gas was released by collision
processes resulting from fission and above 1000°C release was primarily by
diffusion. The two collision processes are (i) "recoil", in which a fission
event near the surface produces a gas atom which has sufficient energy to
escape from the solid, and (ii) "knock-out", when a fission fragment collides

with a gas atom with sufficient energy to knock it out of the solid. These processes have been studied in some detail and formulae have been developed describing the fraction of fission gases released by the processes. However, the fraction is below 1% of the total gas generated, so we may neglect this contribution in first order.

Between 1000° and 1600°C gas is released by diffusion, which includes both atomic diffusion through the lattice and contributions from grain boundaries, cracks and pores.

It was fashionable in the 1960s to derive values of the diffusion coefficient D or to quantify the fractional release f [16]. The analysis was made on a model of the fuel consisting of a number of spheres of radius a. Then

$$f = 6\left(\frac{Dt}{\pi a^2}\right)^{1/2} - 3\,\frac{Dt}{a^2} \qquad (3.4)$$

where t = time of the experiment or irradiation.

We may neglect the second term on the right-hand side. The value of a is usually unknown, and it became common practice to characterize UO_2 by an apparent diffusion coefficient D'.

$$D/a^2 = D'.$$

Hence
$$f = 6\left(\frac{D't}{\pi}\right)^{1/2}. \qquad (3.5)$$

Measurement of D' became a means of characterizing batches or types of UO_2 samples.

However, it became apparent that this method of predicting gas release was not reliable because of the existence of trapping processes, i.e. artifacts that hindered diffusion of the gases [17]. Furthermore, sensitive in-reactor measurements of fission gas pressure, notably by Notley at Chalk River Nuclear Laboratory [18], showed sudden changes in gas release with changes in reactor power that could not be accounted for by diffusion processes. More recent analyses of fission gas behaviour have been based on the direct observations in the electron microscope of fission gas bubbles in UO_2 [19].

A scientifically important discovery was that fission events could cause fission gas bubbles to redissolve into the lattice. Resolution, as it is called, is obviously dependent on the fission rate and it imposes a constant reversal effect on gas bubble growth. Furthermore, since single gas atoms are more mobile than bubbles, it causes increased gas atom mobility — and hence release by single atom diffusion. Technologically the process is not very important. It has a modest effect on gas release in the lower temperature regions of the fuel, where the release rate is very low anyway. High-temperature gas release processes are sufficiently rapid (because diffusion is very rapid) to swamp any resolution effects and they can be ignored by designers.

The original mechanistic study of gas release was made by Barnes and Nelson [20] who calculated the forces exerted by traps on bubbles and derived the sizes to which bubbles had to grow in order to overcome the trapping forces and continue to migrate under various driving forces — thermal gradients, stress gradients, dislocations and grain boundaries. Their analysis is shown schematically in Fig. 3.13. In simple terms, after their

formation, gas atoms: (a) nucleate and grow and move by a surface diffusion
mechanism, (b) are trapped by dislocations that can drag the bubbles, but as
the bubbles grow they drag the dislocations along until (c) grain boundaries
are reached. If grain growth is occurring, the boundaries will drag the
bubbles. However, bubble growth will continue until the bubble drags the
boundary or it escapes and moves towards the fuel centre, probably helping
the sinter pores to form columnar grains [21].

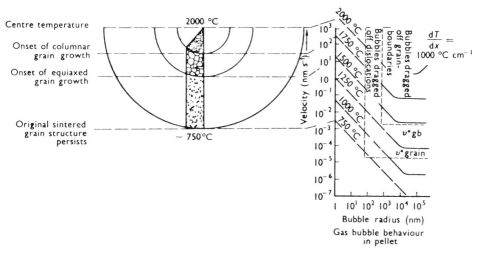

Fig. 3.13. Schematic view of highly rated UO$_2$ pellet.
Source: B. R. T. Frost and B. T. Bradbury.

Stage (c) in this analysis is important. In a fuel pellet whose centre
temperature is 1600°C or less, bubbles will collect on the grain boundaries
in ever-increasing number and size until they touch (Fig. 3.14), *or* until a
reactor shutdown occurs when thermal stresses will cause cracking along the
weakest paths, i.e. the grain boundaries decorated with bubbles. This is the
probable explanation for Notley's observations on gas "bursts" at start-up
and shutdown. A model for this mechanism of gas release has been developed
by Collins and Hargreaves in the UK [22].

While this type of analysis is enlightening to the scientist, it does not
provide a design tool for calculating fission gas release. This is done
through computer models, of which there are two in common use: BUBL and GRASS.
These are time-dependent statistical models which use the Barnes and Nelson
model as a basis, but simplify the mechanisms in order to keep the
calculations within manageable bounds.

In BUBL [23] the fuel is represented by an array of cubic grains in each of
which is a population of dislocations. Gas atoms enter the grains as small
bubbles of fixed size. They stick on the dislocations and grow by agglomeration until one is large enough to move to the grain boundary, where a similar
agglomeration process occurs. BUBL has been refined several times and can
predict fuel swelling due to gas bubbles very reasonably, and can simulate
gas release by escape from grain edges, but does not consider sweeping
mechanisms, such as columnar grain growth.

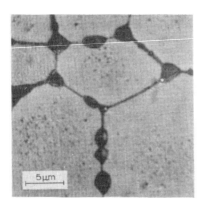 

Fig. 3.14. The development of grain boundary bubbles in
            UO$_2$ under irradiation. Source: D. Collins
            and R. Hargreaves, Physical Metallurgy of
            Reactor Fuel Elements, The Metals Society,
            1975.

The GRASS code [24] uses a distribution function approach. It computes the
evolution of the bubble-size distribution as a function of position in the
fuel for three types of bubbles: those not pinned to structural defects;
those pinned to dislocations; and those pinned to grain boundaries. Bubble
mobility and velocity are computed in terms of surface diffusion, volume
diffusion and evaporation-condensation mechanisms. For each fuel region the
mechanism that gives the largest mobility or velocity is used. The
probability of bubble coalescence is computed. The bubbles are divided up
into equal size ranges on a logarithmic scale and differential equations are
written that describe the rate of change of the number of bubbles in every
size range. Then fuel swelling ΔV is given by:

$$\Delta V = \sum_{\substack{\text{bubble} \\ \text{types}}} \cdot \sum_{\substack{\text{size} \\ \text{ranges}}} \cdot f_i \ (T) \cdot \frac{4}{3}\pi r_i^3$$

where f_i is the number of bubbles in the ith size range and r_i is the average radius of bubbles in the ith size range. Swelling calculations are done separately for each fuel zone. Gas release GR is given by:

$$GR = A \times \sum_{\substack{\text{bubble} \\ \text{type}}} \cdot \sum_{\substack{\text{size} \\ \text{range}}} \cdot f_i \times v_i \times S_i$$

where A is the area of the inner boundary of the columnar grain region, v_i is the average velocity of bubbles in the ith size range, and S_i is the average number of atoms per bubble in the ith range.

GRASS has been extended to GRASS-SST [25] which includes models for intra- and intergranular bubble behaviour and describes the role of interlinked grain boundary bubbles and porosity in gas release and swelling. Furthermore, it deals with transient behaviour of fuel that is valuable in the analysis loss of coolant accidents (LOCA) in light water reactors.

We shall return to these and other models in a later chapter. BUBL, GRASS and GRASS-SST form subroutines of larger fuel element performance codes such as CYGRO and LIFE, but have intrinsic value where gas release and swelling are of special importance.

Fission gases contribute to the swelling of fuel. We may lump them together with all the other fission products to calculate a minimum swelling rate that is based on the atomic volumes of the elements, i.e. one assumes that all the fission products are present as unagglomerated, uncombined atoms. This "base" level is meaningless, however, because fission products do combine and agglomerate. The contribution of fission gas bubbles to fuel swelling is important. If one can calculate the bubble size distribution at any given time by BUBL or GRASS, then one can express this as a distribution function:

$N(R)dR$ = Number of bubbles per unit of total volume with radii in the range R to $R + dT$.

Then the total bubble density is:

$$N = \int_0^\infty N(R)dR$$

and the volume swelling is:

$$\frac{\Delta V}{V} = \frac{4\pi}{3} \int_0^\infty R^3 \ N(R)dR.$$

While resolution was stated to be fairly unimportant in affecting gas release, it may have a more significant effect on fuel swelling by keeping the bubble sizes down.

Let us now consider the role of the other fission products in oxide fuels. For convenience of discussion we will discuss in terms of the classes described above:

1. The noble metals that are present as alloys in the fuel. The nominal composition of the alloy commonly found in the central void is 20% Mo, 17% Tc, 48% Ru, 13% Rh and 2% Pd.
2. Noble metals present in an unalloyed state: Mo, Ru, Tc, Pd, Rd and Ag. These generally occur as white inclusions in the columnar and equiaxed grains (Fig. 3.15). Mo may form MoO_2 or MoO_3 in a region of high oxygen potential and may be used as indicators of local oxygen potentials [26].
3. BaO and SrO which occur as grey inclusions often containing cerium or strontium (see Fig. 3.16).
4. The rare earths and zirconium that form solid solutions in the UO_2 lattice; these actually produce a negative contribution to fuel swelling in UO_2.
5. The volatile elements Rb, Cs, I, Te, Sb and Cd, all of which have high vapour pressures at the fuel operating temperatures and are the most technologically important group of fission products after xenon and krypton.

We are interested in these fission products for two reasons: first, they contribute to fuel swelling. Estimates of the swelling vary between 0.2% and 0.7% $\Delta V/V$ per 1 at % burn-up, the variation being dependent on the oxidation state of the fuel [27]. Clearly xenon and krypton contribute as much or more to the swelling, depending on their bubble sizes as discussed above. In fuel element technology one often measures an overall or integrated swelling that takes account of the temperature gradients in the fuel, as illustrated in Fig. 3.17.

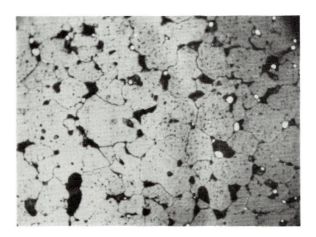

Fig. 3.15. Agglomeration of noble fission products as seen through a microscope.
Source: B. R. T. Frost, R.I.C. Reviews.

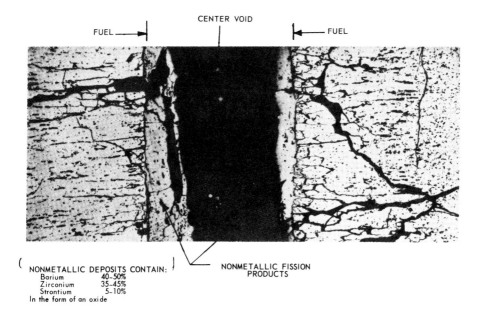

Fig. 3.16.   Typical nonmetallic fission product oxide
deposits (∿100,000 MWd/MT).   Source:
R. Duncan *et al.*, Fast Reactor Fuel Element
Technology, ANS, 1971.

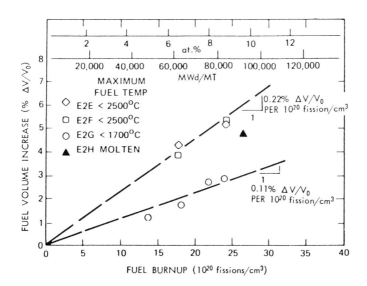

Fig. 3.17.   Effect of temperature on swelling of mixed-
oxide fuel.   Source: R. Duncan *et al.*, Fast
Reactor Fuel Element Technology, ANS, 1971.

A second reason for interest in the solid fission products is the "corrosive" effect on the volatile fission products. In LWR fuels the phenomenon of pellet-cladding interaction (or PCI) is the primary cause of cladding failure [28]. This process is partly mechanical and partly chemical. The mechanical part is attributable to the differential expansion of fuel and cladding during reactor start-ups and shutdowns and to fuel swelling late in life. Its effects may be enhanced locally by pellet-pellet interfaces and by cracks in the fuel. At these locations volatile fission products have easy access to the cladding, so that the process is thought to be a form of stress-corrosion cracking. There is mounting evidence that iodine causes local stress-corrosion cracking at the points of high stress. There is some disagreement over which is the chemical agent among the fission products. This topic will be examined again in Chapter 8, where water reactor fuels are examined in more detail.

In fast reactor oxide fuels large temperature gradients exist from centre to surface, providing a strong driving force for the migration of volatile fission products. It was observed that the stainless steel cladding in LMFBR fuel elements was attacked on its inner wall [28] (see Fig. 3.18). Careful analysis of the fuel and cladding chemistry with electron and ion microprobes showed that caesium migrated to the cladding, bringing with it excess oxygen generated by the fission of uranium in UO_2 or $(U,Pu)O_2$ [8]. Either Cs_2O or Cs_2MoO_4 reacted with the protective oxide film on the stainless steel and proceeded to oxidize the steel along grain boundaries. The nature and rate of attack was temperature dependent (see Fig. 3.19). This process may be cured in at least two different ways: the starting O/M ratio of the fuel may be lowered so that excess oxygen is not available to migrate to the cladding, or one can use an oxygen getter coating on the inner wall of the cladding, e.g. vapour-deposited titanium or zirconium. The former solution works in the hotter regions of the cladding, but then another phenomenon occurs — caesium migrates to the cooler axial blanket region and reacts with the UO_2 blanket to form Cs_2UO_4, which is less dense than UO_2 and so expands and cracks the cladding. The use of getters is being tested.

Another cure for this corrosion effect is to change to carbide fuel, since carbon does not move through the fuel via the volatile fission products and the protective film on the steel remains intact. Furthermore, a UC blanket pellet does not form a compound with caesium.

CARBIDES AND NITRIDES

Some of the physical properties of carbide and nitride fuels were discussed in the previous chapter. Their chemical reactivity with water and oxidizing gases makes them suitable only as LMFBR fuels and in coated particle fuels for HTGRs. As fast reactor fuels the monocarbide (U,Pu)C and the mononitride (U,Pu)N differ in one important respect from $(U,Pu)O_2$; their thermal conductivities are much higher. Hence, for a given fission rate and fuel geometry, carbide and nitride centre temperatures are much lower than for oxide — typically $1400^{\circ}C$ versus $2400^{\circ}C$. The thermal gradients are correspondingly lower. This has a marked effect on fuel behaviour and on fission product distribution [29].

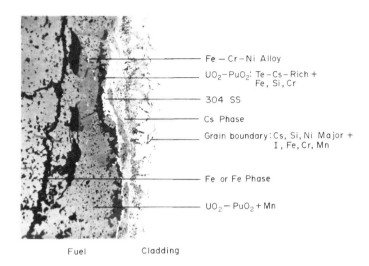

Fuel   Cladding

Fig. 3.18.   Combination attack in Type 304 SS, 4.6 at.%
             burn-up, ∿8 mils of attack, 200X.
             Source: J. D. B. Lambert *et al*., Fast Reactor
             Fuel Element Technology, ANS, 1971.

First, carbide and nitride do not restructure — the final grain size and
shape is similar to the initial grain size and shape. Grain interior pores
do appear to migrate to the grain boundaries, as do the fission gases. A
typical microstructure of carbide or nitride after irradiation to a high
burn-up is shown in Fig. 3.20. The grain boundaires are "decorated" with
fission gas bubbles which contribute to swelling but which only contribute
to gas release if thermal stress cracking links the bubbles. However,
thermal stresses are much lower than in oxide fuel, so that cracking
susceptibility is lower.

The volatile fission products are less mobile than in oxide fuel. Moreover,
the carbides and nitrides of the fission products do not appear to react
with cladding materials in the manner exhibited by the oxides. Thus high gas
release and "fuel"-cladding interactions are absent. On the other hand, the
higher densities of carbide and nitride combined with the retention of
fission products produce a higher swelling rate — by as much as a factor of
two over oxide (Fig. 3.21). Hence, fuel element designs must allow for
accommodation of this swelling. This is usually accomplished through the
use of a "wide" (∿10 μ) sodium bond between the fuel and cladding. However,
carbide and nitride fuel pellets crack under thermal stresses in a different
manner from oxides. They crack into a few large "chunks" — as shown in
Fig. 3.22. In fuel elements that have a lot of free internal volume to
accommodate swelling (sodium-bonded), these chunks may move around and form
a "log jam", i.e. a tight bridge which can transmit swelling forces to local
regions of cladding and cause failure. Collapsible spacers between fuel and
cladding may avoid this problem.

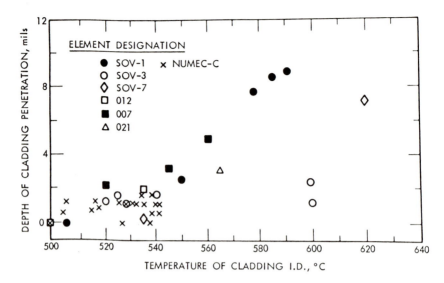

Fig. 3.19. Summary of all ANL results on the depth of
penetration of fission products in the Types
304L and 316 stainless–steel cladding on
EBR-II irradiated mixed-oxide elements.
<u>Source</u>: J. D. B. Lambert, Fast Reactor Fuel
Element Technology, ANS, 1971.

DISPERSION FUELS

In dispersion fuels the fissile phase is surrounded by a high thermal
conductivity matrix and is small enough that temperature differences between
the centre and surface of the fissile phase are small. Hence such fuels
tend to behave as isothermal, heavily restrained pieces of fuel. In most
cases the temperature is such that all fission products are retained within
the fuel particle which is so restrained within the fuel particle that it
cannot swell to any extent. In the HTGR coated particles the temperature
is high enough for volatile fission products to migrate — this accounts for
the need for the silicon carbide layer in the coating which acts as a
diffusion barrier. These particles are usually designed with a low-density
"buffer" layer that helps to accommodate swelling and to absorb the fission
recoil damage. Around any dispersed fissile phase the fission products will
penetrate 10–20 μm into the surrounding matrix and cause severe damage.
Obviously the interparticle spacing must exceed twice the recoil range so
that relatively undamaged matrix remains to give strength and dimensional
stability.

Cermets of UO_2 in stainless steel have been irradiated to very high burn-ups
of the UO_2 phase without failure. The composition and structure of the UO_2
phase has not been well examined, but it must differ greatly from the original
unirradiated material. It is fully dense, probably noncrystalline and
contains a lot of fission products, yet it retains its original geometry.

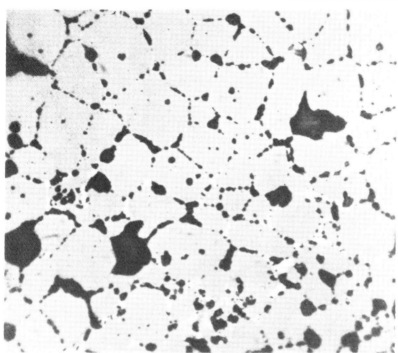

Fig. 3.20. Grain boundary bubbles in irradiated mixed
carbide (500X). <u>Source</u>: B. Harbourne *et al.*,
Fast Reactor Fuel Element Technology, ANS, 1971.

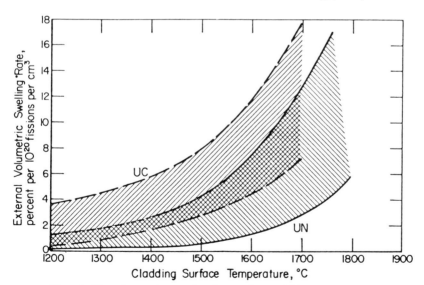

Fig. 3.21. Comparison of volumetric swelling rate as a
function of cladding surface temperature for
UC and UN clad in W-25 wt.% Re.
<u>Source</u>: D. Hilbert *et al.*, ANS, 1971.

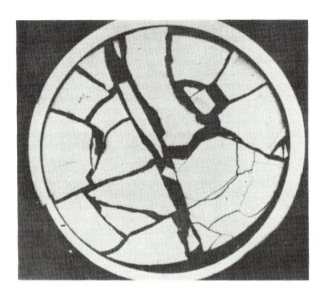

Fig. 3.22.   Cracking of mixed carbide fuel.   <u>Source:</u>
A. Bauer *et al.*, ANS, 1971.

REFERENCES

1. Glasstone, S. and Sesonske, A., *Nuclear Reactor Engineering*, Van Nostrand Reinhold, New York (1963).
2. Billington, D. and Crawford, J. H., *Radiation Damage in Solids*, Princeton University Press, Princeton, NJ (1961).
3. Bradbury, B. T. and Frost, B. R. T., *Studies in Radiation Effects in Solids*, Vol. 11, p. 159 (ed. G. J. Dienes), Gordon & Breach, New York (1967).
4. Dienes, G. J. and Vineyard, G. H., *Radiation Damage in Solids*, pp. 830-860, Academic Press, New York (1962).
5. Davies, J. H., unpublished work at Atomic Energy Research Establishment, Harwell, England.
6. Wait, E. and Frost, B. R. T., *Proceedings of IAEA Conf. on Plutonium as a Reactor Fuel*, paper SM 88/25, p. 469, Brussels (1967).
7. Anselin, F. and Baily, W. E., *Trans. ANS*, 10, 103 (1967).
8. IAEA: *Proceedings of a Panel on the Behavior and Chemical State of Irradiated Ceramic Fuels*, IAEA, Vienna (1974).
9. Gittus, J. H., *Uranium*, Butterworths, Washington (1963).
10. Pugh, S. F. and Butcher, B. R., *Reactor Technology--Selected Reviews 1964*, USAEC, p. 331, "Metallic fuels".
11. Murphy, W. F., Beck, W. N., Brown, F. L., Koprowski, B. J. and Neimark, L. A., Argonne National Laboratory Report ANL-7602 (1969).
12. Nichols, F. A., *J. Nucl. Mat.* 84 (Books 1, 2), 1-25 (October 1979).
13. Report on Densification by USAEC Regulatory Staff, WASH-1236 (Nov. 14, 1972).
14. Meyer, R. O., *The Analysis of Fuel Densification*, NUREG-0085 (July 1976).
15. Haneirk, A., Arnesen, P. and Knudsen, K. D., BNES Conference on Nuclear Fuel Performance, London, paper No. 89 (1973).

16. Booth, A. H., Canadian Report CRDC-721 (1957).
17. MacEwan, J. R. and Stevens, W. H., *J. Nucl. Mat.* 11, 77 (1964).
18. Notley, M. F., DesHaies, R. and MacEwan, J. R. Atomic Energy of Canada Limited Report AECL-2662 (1966).
19. Speight, M. V., *Physical Metallurgy of Reactor Fuel Elements*, The Metals Society, pp. 222-230 (1975).
20. Barnes, R. S. and Nelson, R. S., U.K. Atomic Energy Authority Report R4952 and AIME Nuclear Metallurgy Symposium on Radiation Effects in Solids, Asheville, NC (1965).
21. For a summary of fission gas release models see: Frost, B. R. T., *Nuclear Applications and Technology*, 9 (2), 128 (1970); and Olander, D., *Fundamental Aspects of Nuclear Reactor Fuel Elements*, DOE-TIC, p. 287 (1976).
22. Collins, D. A. and Hargreaves, R., *Physical Metallurgy of Reactor Fuel Elements*, The Metals Society, pp. 253-258 (1975).
23. Warner, J. R. and Nichols, F. A., *Nuclear Appl. & Tech.* 9 (2), 148 (1970).
24. Poeppel, R. B., *Proceedings of a Conference on Fast Reactor Fuel Element Technology*, ANS, p. 311 (1971).
25. Rest, J., NUREG/CR-0202; Argonne National Laboratory Report ANL-78-53 (June 1978).
26. Johnson, C. E., Johnson, I. and Crouthamel, C. E., *Proceedings of a Conference on Fast Reactor Fuel Element Technology*, ANS, p. 393 (1971).
27. Anselin, F. and Baily, W. E., *Trans. ANS* 10, 103 (1967).
28. A number of papers on the P.C.I. phenomenon appear in the *Proceedings of a Conference on Water Reactor Fuel Performance*, ANS (1977).
29. Lambert, J. D. B., Neimark, L. A., Murphy, W. F. and Dickerman, C. E., *Proceedings of a Conference on Fast Reactor Fuel Element Technology*, ANS, p. 517 (1971).
30. *Proceedings of a Conference on Fast Reactor Fuel Element Technology*, ANS (1971); see papers in Session VI, Performance of Advanced Fuels, pp. 741-894.

APPENDIX. BURN-UP AND RATING OF NUCLEAR FUELS

In the studies of irradiation effects in nuclear fuels confusion frequently arises from the different units which are used by different groups of workers to express the number of fission events which have occurred in any experiment. The confusion arises from two main sources:

(i) Those accustomed to radiation damage studies in nonfissile materials tend to express damage in fuels in terms of neutrons/cm^2. Since fission fragments are primarily responsible for the damage in fuels, a neutron dose in isolation does not convey much information. The number of fissions occurring in a given neutron dose is proportional to the density of fissile atoms and to their fission cross-section, in the particular neutron energy spectrum to which they are exposed. Thus, if we know the fuel composition and density, the degree of enrichment of uranium atoms and the neutron spectrum we can calculate numbers of fission events from a neutron dose. It should be remembered that in any reactor the neutron intensities and spectra can vary across the core, and thus there will be an axial variation in burn-up. Also the high absorption cross-sections of the fissile and fertile isotopes can produce self-shielding effects, so that the burn-up can vary from the surface to the centre of a fuel specimen.

(ii) The densities of fissile compounds vary over a wide range, and when we consider also dispersions in non-fissile materials it becomes difficult to compare the relative effects of a given amount of damage in several different materials. A convenient answer to this problem is to express burn-ups as fissions/cm^3 which is the most logical expression for damage.

Those who are interested in the practical performance of nuclear fuels quote "damage" in different terms: the more obvious is the burn-up in terms of the percentage of heavy atoms which have fissioned, the more common is the net thermal power generated per metric ton of fuel (megawatt-days/tonne). The conversion factors for these units are as follows:

Material	Fissions/cm^3	% burn-up of heavy atoms		MWd/t[*]	
Uranium	10^{20}	0.209		1810	
	4.78×10^{20}	1.0		8650	
			U		UO_2
UO_2	10^{20}	0.411	3560		3130
	2.43×10^{20}	1.0	8650		7630
					UC
UC	10^{20}	0.305	2640		2500
	3.28×10^{20}	1.0	8650		8220

These values are for theoretically dense materials.

A further unit of interest in practice is the "rating" or the rate at which the fuel is burned up. This is generally expressed as watts/g (total fuel or fissile isotope) or MW/t (which is the same thing) and sometimes as watts/cm^3 or MW/litre. The linear rating is expressed as kW/m or kW/ft.

Radiation Damage

Radiation damage is generally expressed in terms of the number of fast neutrons/cm^2 of irradiated material. A common designation is the number of neutrons with energies greater than 0.1 MeV that strike a cm^2 of sample, e.g. 10^{22}n (>0.1 MeV/cm^2). This is derived either by monitoring the neutron source to determine the flux >0.1 MeV and multiplying by the exposure time in seconds *or* by exposing suitable activation foils to the flux and measuring the induced activity. By careful selection one can find foils with different energy thresholds for reactions and hence derive a spectrum.

Because ion simulation of neutron damage has become commonplace, a unit has been named that can describe damage regardless of its origin. Displacements per atom or dpa refers to the number of times that an atom is displaced from its lattice site by displacement or replacement collisions. It is calculated from collision theory. For perspective, 10^{23} n/cm^2 of neutrons >0.1 MeV is equivalent to $\sim$60 dpa.

---

[*]In more recent publications burn-up may be expressed as MW/kg, which is 10^{-3} times MW/t. t is usually a metric ton = 2200 lb = 1000 kg.

CHAPTER 4

Cladding and Duct Materials

DESIGN CONSIDERATIONS

The functions of the cladding material are to maintain a barrier between the fuel and the coolant and to maintain a predetermined geometry of the fuel element. The maintenance of a barrier between fuel and coolant is the more important function; the cladding is generally the only feature that stops the fuel and fission products from getting into the primary circuit. Most reactors are operated under the philosophy of keeping the primary circuit as clean as possible so that maintenance is easier and the potential for spread of radioactivity into the reactor shell and beyond is minimized.

The design and development of fuel elements will be discussed in the following chapter. However, the reader must be reminded at this juncture that the reactor core is assembled from a number of fuel element bundles or sub-assemblies, each of which contains a large number of fuel rods or elements. Thus, in addition to the fuel tube (cladding) containing the fuel, the sub-assembly contains spacers to provide the desired coolant channel size and geometry, and ducts to provide strength and to direct coolant flow. Spacers and ducts must not break or distort in service, and we must include them in our discussion of the non-fuel components of the core.

We must anticipate the next chapter to some extent in order to define the size, shape and material of the cladding and ducts. These considerations will vary between fast and thermal neutron spectra and between coolants and between fuels.

In general, the cladding and duct materials should have the lowest possible neutron absorption cross-sections. Table 4.1 lists the thermal neutron absorption cross-sections for a number of metals for comparative purposes. With a practical upper value of ∿0.2 barns, one is limited to aluminium, magnesium, zirconium, beryllium and graphite* for thermal reactors, although *thin* stainless steel has been used in the British AGRs. On the other hand, stainless steels are used extensively in fast reactors, where the cross-sections are lower, and vanadium alloys have been seriously considered, but

---

*Not listed in Table 4.1; its value is comparable to beryllium.

the other refractory metals (molybdenum, niobium and tantalum) have high resonance cross-sections in the keV neutron energy range and are unacceptable. One must exercise careful control over the amounts of minor elements that have high absorption cross-sections, e.g. boron in stainless steel and hafnium in zirconium.

TABLE 4.1.   Selected Values of Thermal Neutron
Absorption Cross-sections

Element	Barns
Aluminium	0.23
Beryllium	0.010
Chromium	2.9
Copper	3.7
Iron	2.5
Magnesium	0.063
Molybdenum	2.5
Nickel	4.6
Niobium	1.1
Tantalum	21.0
Titanium	5.6
Tungsten	19.0
Zirconium	0.18

Source: B. R. T. Frost and M. Waldron.

The cladding in a conventional rod-type fuel element is subjected to steadily rising stresses due to fuel swelling and fission gas release, to a steady coolant pressure on the outside, to thermal stresses that vary with power, and to local, fluctuating stresses at cracks and pellet-pellet interfaces in the fuel column. To withstand these stresses the cladding must possess and retain good multiaxial rupture strength, creep strength and ductility. The wall thickness of the cladding should be as thin as possible, consistent with adequate mechanical performance, for reasons of good neutron economy and minimum thermal stresses. On the other hand, the fuel and coolant may react slowly with the cladding so that a "wastage" allowance must be made. These reactions may also affect the mechanical properties (as in hydriding of Zircaloy and carburizing of stainless steels) and will have to be taken into consideration in safety analyses. In such analyses the question is often asked: can the fuel and cladding react to form low melting-point phases? This is certainly the case for uranium metal and stainless steel, where low melting-point eutectics can form (see Appendix to Chapter 2).

Obviously, the design of fuel element cladding is an exercise in optimization or a series of compromises. Simplistically, one calculates the mechanical properties required to withstand the postulated stresses, selects a low cross-section material with the correct properties of strength and ductility and then factors in an appropriate wastage allowance to arrive at a wall thickness. Heat transfer and fuel performance considerations will determine the tube (fuel) diameter and it is important to remember that the stress on a tube (ρ) is related to the ratio of its wall thickness (t) to its diameter (d) by the relationship

$$\rho = \frac{p \cdot d}{t}$$

where p – the coolant pressure.

Obviously ρ must not exceed the allowable stress, e.g. stress for 1%
deformation in 10^4 hours. The cladding design optimization is greatly helped
by fuel element modelling, which will be discussed in the following chapter.
Models require as input a number of properties of cladding in the irradiated
and unirradiated state.

Heat transfer considerations may require that the outer surface of the tube
be roughened. This is necessary in gas-cooled reactors to promote turbulence
or reduce the effective thickness of the boundary layer and avoid excessive
cladding temperatures. Roughening can be in the form of machined-on small
fins, as in the stainless steel cladding of the gas-cooled-fast reactor (GCFR)
and the advanced gas-cooled reactor (AGR), or spiral fins extruded on aluminium
or magnox in CO_2-cooled or air-cooled thermal reactors shown in Fig. 4.1. In
general, these fins enhance the strength of the cladding, but of course they
contribute to neutron absorption and may bear the brunt of the coolant
corrosion.

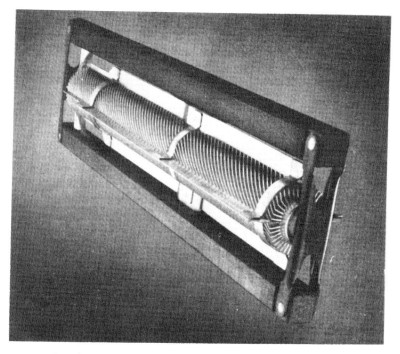

Fig. 4.1.   Improved design of Berkeley fuel element.
Courtesy: UKAEA.

Ducts are generally massive boxes of square or hexagonal cross-section, made
by extrusion or by bending and welding sheet (see Fig. 4.2, item 5). Fuel
element spacers may be wires wrapped around the elements in a spiral and
welded to each end, or grids are made up from welded metal strip and slipped
over the plain tubular elements, their outer boundaries being welded to the
duct.

GCFR FUEL ASSEMBLY

1. *Exit Nozzle*
2. *Top Load Pad*
3. *Replacement Orifice*
4. *Exit Shielding*
5. *Flow Duct*
6. *Back-up Support Grid*
7. *Fuel Rod Spacer Grid*
8. *Above Core Load Pad*
9. *Fuel Rods*
10. *Grid Manifold*
11. *Grid Plate Shielding*
12. *Pressure Equalization Vent*
13. *Fission Product Trap*
14. *Inlet Nozzle*
15. *Seal Rings*

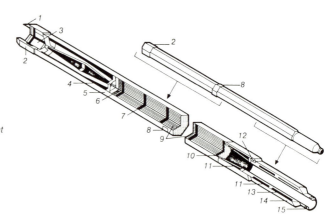

Fig. 4.2. GCFR fuel assembly. <u>Courtesy</u>: General Atomics.

COMMONLY USED MATERIALS AND THEIR PROPERTIES

Aluminium and its alloys are used extensively as the fuel matrix and as cladding in research reactor fuel elements, which are generally of plate or hollow cylinder geometry. The temperature of the coolant in these reactors is generally around 100°C or lower so that corrosion is not a serious problem. The mechanical strength requirements of the alloy are not arduous provided the element is well made and the fuel "meat" and cladding are fully bonded (this will be discussed later).

A common alloy that is used in research and test reactor fuel elements is 6061 which contains 1% Mg, 0.6% Si, 0.25% Ca and 0.25% Cr. It has good corrosion resistance and is easily fabricated.

Magnox. Magnox AL80 is a magnesium-base alloy containing about 0.8% Al, 0.002–0.05% Be, 0.008% Ca and 0.006% Fe [1]. The alloying elements provide good corrosion resistance to the CO_2 coolant in the Calder Hall and Magnox type (e.g. Berkeley) reactors which use adjusted uranium fuel. This cladding is rather weak but very ductile. The metal fuel provides most of the structural strength of the bar (see Fig. 4.1). The cladding is pressurized onto the fuel at 500°C to provide a good thermal bond. Fuel swelling and growth in the desired burn-up range (up to ∿5000 MWd/te) is relatively low and the cladding is sufficiently ductile to adjust to fuel movement. Cladding failure results in oxidation of the uranium and severe local distortion or blistering. The performance of Magnox fuel elements will be discussed in a later chapter.

Zirconium. All of the light-water and heavy-water moderated reactors use zirconium alloy cladding and the heavy-water reactors use zirconium alloy pressure tubes. There is a voluminous literature on these alloys and their development is one of the success stories of nuclear metallurgy [2,3].

Zirconium has very attractive nuclear properties once it is separated from hafnium, with which it generally occurs in nature. The Van Arkel iodine refinement process and the Kroll process led to the separation of pure zirconium and to the application of this material to pressurized water reactors.

Zirconium is a hexagonal metal and when drawn into tubes it exhibits a
marked texture, i.e. a preferred orientation of crystal planes, that lead to
directionality of properties that cannot be eradicated by heat treatment.
Hence it is essential to determine the properties in terms of tube
orientation.

Pure zirconium in high pressure water at ~300°C (typical of BWR and PWR
conditions) gains weight due to surface oxidation, forming a thin protective
black film [4]. The weight gain follows a curve that is somewhere between
a parabolic and a cubic law dependence (Fig. 4.3). However, after a certain
time the rate changes to a linear dependence (breakaway corrosion) due to
film cracking. White oxide spots appear and the film cracks. This phenomenon
is associated with certain impurities such as nitrogen, carbon, oxygen,
aluminium, titanium and silicon.

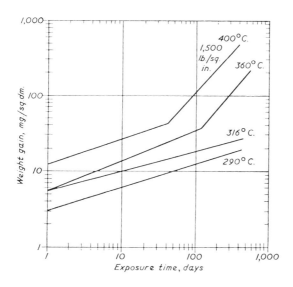

Fig. 4.3. Weight gain versus time for zirconium base
alloys. Source: B. R. T. Frost.

To overcome breakaway corrosion a series of alloys, called Zircaloys, were
developed. The compositions of these alloys are given in Table 4.2. The
two which are in common use are Zircaloy-2 and Zircaloy-4. The additions of
iron, chromium, nickel and tin combined with the restriction of the nitrogen
content reduce the tendency for film cracking. Zircaloy-4 offers better
resistance to hydrogen uptake than Zircaloy-2.

Hydrogen uptake is a potential problem in zirconium because the metal has a
strong affinity for hydrogen. The solubility limit is 75 ppm at 300°C; above
this level hydride precipitates out as plates. These plates orient themselves
according to the texture of the metal. Since the hydride phase is brittle,
there is a large variation in strength and ductility with texture.
Figure 4.4. shows a hydrided Zircaloy tube with circumferentially oriented
hydride plates — a favourable orientation.

TABLE 4.2. Compositions of the Zircaloy Series of Alloys

Alloying addition (wt%)	Zircaloy 1	Zircaloy 2	Zircaloy-3			Nickel-free Zircaloy-2	Zircaloy 4
			a	b	c		
Sn	2.5	1.50(1.20-1.70)	0.25	0.50	0.50	1.50(1.20-1.70)	1.50(1.20-1.70)
Fe	—	0.12(0.07-0.20)	0.25	0.40	0.20	0.15(0.12-0.18)	0.20(0.18-0.24)
Cr	—	0.10(0.05-0.15)	—	—	—	0.10(0.05-0.15)	0.10(0.70-0.13)
Ni	—	0.05(0.03-0.08)	—	—	0.20	<0.007	<0.007
Total Fe, Cr, Ni	—	(0.18-0.38)	—	—	—	—	>0.28
O(ppm)	—	1000-1400				1000-1400	1000-1400

Maximum impurity levels:

Element:	Al	B	Cd	C	Co	Cu	Hf	H	Mn	N	Si	Ti	W	U
ppm:	75	0.5	0.5	270	20	50	200	25	50	80	200	50	100	3.5

Source: B. Cox, *Advances in Corrosion Science*, ed. Fontana and Staehle.

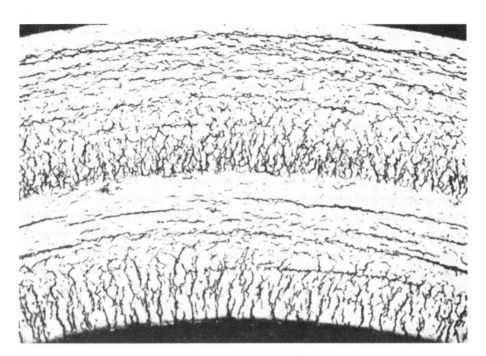

Fig. 4.4. Zirconium hydride in cladding tube; circumfer-
ential directionality. <u>Source</u>: D. Pickman,
Physical Metallurgy of Reactor Fuel Elements,
The Metals Society, 1975.

This texture is achieved by using large ratios of wall thickness reduction
to tube diameter reduction during processing. Worries about hydriding have
diminished as experience has accumulated [5]. The hydrogen solubility limit
is exceeded only late in fuel element life and the hydride has some ductility
above $\sim 150^{\circ}$C. Of course, operation under loss-of-coolant (LOCA) conditions
will produce much more hydriding; this regime is discussed in a later chapter.

Stainless steels. Generally, stainless steel cladding is made of Types 304
or 316 steel. These alloys were originally developed for use in superheaters
and in chemical plants. They possess good strength up to $\sim 600^{\circ}$C and are
normally quite ductile. Hence they are attractive for use in reactors that
operate with coolant outlet temperatures in excess of 500°C. However, the
British AGR reactor, which is CO_2 cooled, uses an Fe-20Cr-25Ni-1Nb alloy
which has good corrosion resistance and reasonable strength up to 800°C.
The AGR cladding pioneered the use of double-vacuum melting in nuclear
material. Because of the thin wall, no nonmetallic (slag) inclusions can be
tolerated, so the vacuum melting process was adopted along with a specification
that required a grain size that ensured the presence of several grains across
the wall. This technology has been adopted in other stainless steel
applications, especially breeder reactors.

In the USA, stainless steel cladding was used for light-water reactors for a
while, but has been totally superseded by Zircaloy which has given better
performance.

Stainless steels are used in all current generation fast breeder reactors for all non-fuel core components. Such reactors demand a more arduous fuel element performance: the burn-up target is 100,000-150,000 MWD/te or 10-15 at %, the failure rate must be very low and the radiation environment is severe. These alloys are complex since the Fe-Cr-Ni phase diagram contains some brittle phases (like sigma), that must be avoided, and chromium forms stable carbides that affect mechanical properties. Stainless steels are not always thermally stable; their microstructure changes with time at temperature and its properties change accordingly, i.e. aging is an important phenomenon.

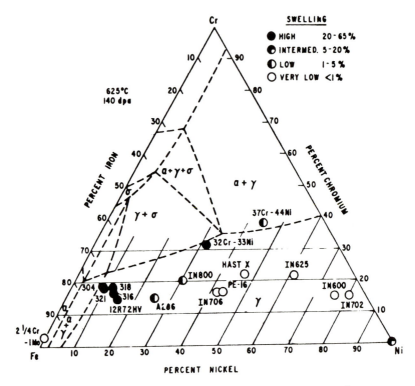

Fig. 4.5.   Iron-chromium-nickel phase diagram at 625°C
             isotherm showing the compositions of commercial
             alloys. <u>Source</u>: B. R. T. Frost.

A number of nickel-base alloys have been developed for use in the power and chemical industries, mostly by INCO [8]. These include Incoloy 600, 625 and 800 and intermediate alloys such as PE-16 in Europe or Incoloy 706 in North America. Alloys such as PE-16 contain aluminium and/or titanium so that γ' phases form (Ni_3Al or Ni_3Ti). By suitable heat treatment these form a finely dispersed second phase that imparts greatly improved high temperature creep strength to the alloys (hence these are generically known as "superalloys").

Both the stainless steels and the nickel-base alloys have the face-centred-cubic gamma (austenite) phase structure which is characteristically ductile and easy to form. Ferritic stainless steels, such as alloy 410, contain no nickel and have the body-centred-cubic alpha (ferrite) phase structure. These alloys possess less ductility, are difficult to fabricate and are not as strong as the austenitic steels at 600°C. Nevertheless, they have some attractions from a radiation damage standpoint.

The properties of stainless steels (austenitic and ferritic) are well described in the *American Society for Metals Handbook*.

Thermal stress (ρ_t) is proportional to the coefficient of thermal expansion α and inversely proportional to the thermal conductivity K:

$$\rho_t = \frac{\alpha E}{(1-\nu)K}$$

where ν = Poisson's ratio,
     E = Young's modulus.

The austenitic steels have a high coefficient of thermal expansion and a low thermal conductivity, which means that they generate high thermal stresses compared to the ferritics or to the refractory metals like niobium and vanadium. They do, however, possess high strength and ductility, have good oxidation-resistance and are easily fabricated. They are also relatively cheap.

COMPATIBILITY

In this section we will discuss cladding-coolant and cladding-fuel compatibility in terms of general principles, because the specifics will vary from reactor to reactor and even from point to point within any given reactor. The classical method of reporting compatibility and corrosion data is in terms of weight gain or weight loss versus time for a given set of conditions. For reactor cladding the effective (load-bearing) wall thickness as a function of time is a better measure, although both measures ignore the important local effects such as stress-enhancement at the tips of cracks or pits and special chemical effects as in stress-corrosion cracking. In the final analysis it is the *effective* strength and ductility that matter in relation to whether or not the cladding will fail.

Zirconium. The corrosion of zirconium by high pressure water has already been discussed because of its influence on the development of the Zircaloys. Zircaloys may also be attacked on the inner wall of the cladding by two different mechanisms. In one case localized hydride formation can occur, e.g. through retained water vapour in the fuel. A typical effect is the "sunburst" which is sometimes seen in the post-irradiation examination of fuel pins (Fig. 4.6). This can be cured by careful quality control during fuel pin manufacture to ensure the removal of hydrogenous materials. In the other case, discussed more fully in a later chapter on water reactor fuel elements, certain fission products (e.g. iodine) may induce local stress corrosion, which combine with the localization of stresses due to discontinuities in the fuel column to cause the cladding to crack. This is generally referred to as pellet-cladding-interaction or PCI and is sensitive to the mode of operation of the reactor.

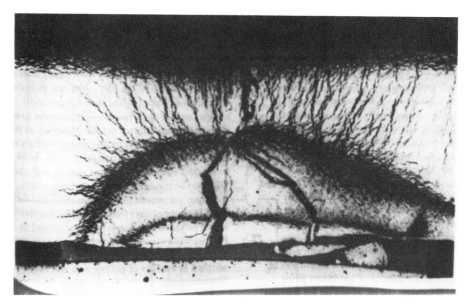

Fig. 4.6. Hydride 'sunbursts' on inner surface of cladding.
Source: D. Pickman, Physical Metallurgy of
Reactor Fuel Elements, The Metals Society, 1975.

Stainless steels. The temperature at which metallic uranium or plutonium can
operate in contact with stainless steel is limited by the formation of low
melting point eutectics of U and Pu with Fe, Cr and Ni. In the uranium case
this is around 725-750°C (Fig. 4.7) and for plutonium it is around 450°C.
The main concern with U-Pu alloys for fast reactor use is the possibility
of forming a liquid phase with stainless steel cladding during a loss-of-
coolant-flow type of transient incident. The addition of zirconium to the
U-Pu fuel raises the reaction temperature or reduces the reaction kinetics
sufficiently for U-Pu-Zr to be a potential fast reactor fuel. Tests at
Argonne National Laboratory [19] showed that a U-15Pu-10Zr alloy is compatible
with Type 304 stainless steel to temperatures approaching 1500°F.

In the unirradiated state neither UO_2 nor $(U,Pu)O_2$ react to any extent with
stainless steel. However, after irradiation up to and beyond 4 a/o burn-up
corrosive-type reactions are observed at the fuel-cladding interface. These
have been studied in considerable detail by means of electron and ion microprobes that can analyse the chemical composition on the micron scale. What
appears to happen is that caesium and molybdenum (as the oxide) migrate to
the fuel-cladding interface and form an electrochemical cell of caesium
molybdate that acts as a medium for transferring oxygen from the fuel to the
cladding [10]. The attack takes several forms, depending on the temperature
and the fuel chemistry. From a performance viewpoint the intergranular
attack is the most weakening. However, there is little hard evidence that
failures occur at the site of this cladding attack although it is expected
that it will enhance the likelihood of failure during a power or coolant
transient.

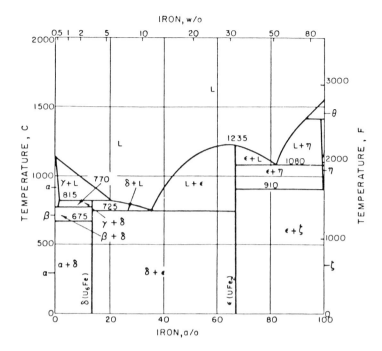

Fig. 4.7.   The uranium–iron phase diagram.   Source:
F. Rough and A. Bauer, BMI-1300.

Remedies for this form of attack are to reduce the initial O/M ratio of the
fuel or to incorporate a getter (such as titanium) inside the fuel cladding.
Carbide fuels do not cause this form of reaction, although another problem
exists in the carburization of stainless steels. Particularly when a sodium
bond is used and the carbide contains dicarbide needles, carburization of
the stainless steel to form embrittling chromium carbides is likely. This
can be offset by fabricating the fuel to be single phase or to contain only
modest amounts of the sesquicarbide phase.

In a similar vein the external surface of the cladding may be carburized by
carbon dissolved in the sodium coolant. Sources of such carbon are not
difficult to find — any ferritic steels in the circuit, oils from pump seals
and even stainless steel at a lower temperature. Given the existence of a
small amount of carbon in solution, there is a driving force to transfer
carbon to the hottest stainless steel in the circuit, i.e. the cladding.
However, provided reasonable precautions are taken to eliminate carbon sources
and the cladding is not run above 650°C, carburization is not a serious
problem. The transport of carbon around a sodium circuit and its uptake by
stainless steel have been studied experimentally and analysed in a model that
incorporated thermodynamic and kinetic data [11].

RADIATION EFFECTS

The unique feature of the environment in which cladding and ducts must operate is radiation. This is made up of neutrons (of varying energy) and gamma rays. In very general terms we may neglect the gamma rays, except for their heating effects, and focus on the radiation damage caused by those neutrons whose incident energy is greater than 0.1 MeV. Clearly the number of such neutrons will depend on the type of reactor under consideration. In a large LWR the peak damage flux in the core is around 10^{12} n/cm^2/sec or less, while in a 1000 MWe fast breeder reactor it would be close to 10^{16} n/cm^2/sec.

In order to understand the effects of fast neutrons on material properties we must first consider their basic reactions with metals [12,13,14]. A very energetic neutron slows down in a solid by ionizing the atoms along its track until it is going slowly enough to collide with an atom, which is then displaced (Fig. 3.3). This primary knock-on atom or PKA moves off and creates additional displacements. The displaced atoms leave behind vacant lattice sites or "vacancies". Eventually these displaced atoms come to rest in the lattice as "interstitial" atoms. Since there is thermal motion in the lattice *and* because the displacements stir up the lattice, some interstitials find vacant sites and cease to be defects (they "recombine"). After a single neutron event the equilibrium configuration is often called thermal spike. Similar effects can be produced by bombarding solids with charged particles, such as protons or nickel ions, although their path lengths in solids are generally much shorter than for neutrons.

Neutron bombardment also causes transmutations to occur. The simplest form is "activation" when the neutron is captured after causing damage and causes a transition of one or more isotopes of the host material to be excited into a radioactive state, producing beta or gamma rays, and a different element. A more complex reaction is the (n, α) or (n, p) reaction which energetic neutrons may have with various isotopes. There is often an energy threshold in the MeV region for such reactions. They are important in that (n, α) reactions result in the formation of helium atoms that can have technological effects.

The practical consequences of radiation damage stem from the production of vacancies, interstitials and helium atoms. Certain properties of materials stem from the concentration and behaviour of lattice defects, that are present in unirradiated materials but in lower concentrations. These properties are related to atomic diffusion which is controlled by the defects. Hence, diffusion-controlled processes such as corrosion and certain modes of creep will be accelerated when radiation increases the defect concentration. This acceleration is dependent on the instantaneous population of defects, i.e. is dependent on the neutron *flux*.

Other radiation effects are cumulative in nature, i.e. they increase as the *fluence* or dose increases. The hardening of metals by radiation is typical and we may use it by way of illustration. The hardness or strength of a metal is determined by the level of difficulty that dislocations[*] experience in moving through the material in response to an applied load or stress. Grain boundaries, dislocations and precipitates act as obstacles to dislocation motion — they harden and strengthen the metal.

---

[*]Dislocations are discontinuities in the crystal lattice that allow deformation processes to occur easily, e.g. slip and shear along crystal planes.

The cladding of a fuel element must possess high strength to resist the stresses caused by fuel swelling and gas release. It must also possess good ductility to accommodate the cumulative stresses and strains that arise during the life of the fuel element in the reactor. Thus, we are very interested in the extent to which irradiation changes strength and ductility. Very crudely, radiation effects are similar to work hardening where the yield strength of metals and alloys increases and the ductility decreases. In work hardening the effects are caused by the build up of dislocation networks, which hinder the passage of other dislocations. In precipitation hardening the small precipitates hinder dislocation motion and in radiation damage defect clusters hinder dislocation motion.

The defect clusters are formed when individual defects migrate and find others of the same type. They accumulate because the free energy of the system is thereby reduced. The most stable configuration of a defect cluster is a flat plate or disc. These discs act like precipitates in stopping the movement of dislocations. Thus the yield stress rises with radiation dose and the ductility or work hardening range decreases. These effects are temperature-dependent because defect clusters grow and finally disappear or anneal out with increasing temperature. The yield strength changes with temperature in a fairly regular fashion but the ductility tends to vary in a fluctuating manner (Figs. 4.8 and 4.9). The figure indicates a softening around 500°C due to thermal annealing of work hardening followed by a combination of defect cluster and helium embrittlement. The defect clusters anneal out around 650°C, but helium bubbles continue to grow on grain boundaries above this temperature, causing low ductility.

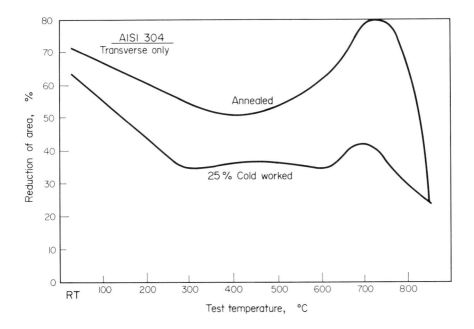

Fig. 4.8.   The effect of test temperature on the ductility
of AISI Types 348 and 304 stainless steel.

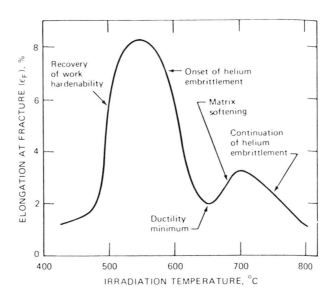

Fig. 4.9.   The effect of irradiation temperature on the
ductility of irradiated stainless steel.
Source: D. Olander, Fundamental Aspects of
Nuclear Fuel Elements TID-26711 (Fig. 18.42).

In 1967 another radiation effect was first observed when stainless steel
irradiated to a high fluence in the Dounreay Fast Reactor was examined in an
electron microscope [15]. The structure was seen to resemble Swiss cheese,
being full of many small holes or "voids" (Fig. 4.10). Subsequent research
has shown that these produce a severe technological problem in the development of fast reactor cores. Voids are formed because of an excess of
vacancies that result from fast neutron bombardment in the temperature range
0.2–0.5 of the melting temperature. Below 0.2 T_m the vacancies are immobile;
above 0.5 T_m they are too mobile — they move rapidly to sinks such as grain
boundaries. In the critical temperature range interstitials are preferentially
absorbed by dislocations, leaving the vacancies free to cluster. Normally
clusters collapse into flat plates, but the helium produced by (n, α)
reactions prevents this. Hence voids or low pressure bubbles are formed and
grow in number and size as the fluence increases. In macroscopic terms, the
metal swells in volume with increasing fluence [16]. The swelling temperature relationships follows a bell-shaped curve, the peak swelling for stainless steels being around 500 C (Fig. 4.11).

To recapitulate, fast neutron irradiation increases the yield stress and
decreases ductility, promotes accelerated creep and produces void swelling.
Let us consider the practical consequences of these phenomena. If one only
examines the fuel element or rod, the consequences may even be favourable,
since the neutron-induced swelling of the cladding may cause it to move away
from the swelling fuel or at least reduce the swelling stresses on the
cladding. However, one must be cognizant of the temperature variation of
void swelling in relation to the fuel swelling.

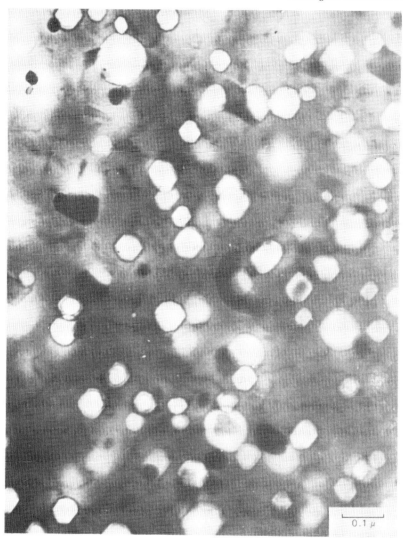

Fig. 4.10. Void structure in irradiated Type 316 stainless
steel. Source: Oak Ridge National Laboratory.

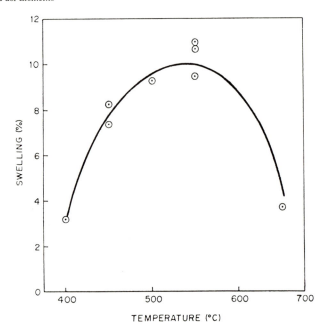

Fig. 4.11.   Temperature dependence of void swelling in
Type 316 stainless steel. <u>Source</u>: Oak Ridge
National Laboratory.

When one examines the fast reactor core assembled from a number of hexagonal
subassemblies the consequences of void swelling become more serious.
Because of neutron leakage the flux drops off across the core towards the
outer edges. A subassembly is sufficiently large that one side may be exposed
to a higher flux (and fluence) than the other, i.e. the side nearer to the
core centre will swell more rapidly than the opposite face, producing a
"bimetallic strip" effect leading to bowing. If allowed to proceed unchecked
this could result in jamming of the core so that unloading became impossible.
This void swelling rather than fuel element performance could place the
upper limit on core life and hence on reactor economics [17].

Several solutions to this problem have been proposed. The simplest is to
rotate the subassemblies periodically to even out the neutron exposure, but
operationally this is not easy. Another approach is to hope that the
radiation-induced creep will help to relax the interference pressures between
subassemblies, but this is uncertain. Finally, new alloys might be developed
with low swelling characteristics; this path is the one most followed worldwide. In general the approach has been Edisonian in character [18].

The earliest efforts were aimed at changing the sink density for defects by
varying the amount of cold work and the grain size. Then changes were made
to the level of minor alloying elements in Type 316 stainless steel; it was
found the ~1% Si and 1% Ti appreciably lowered the swelling rate. This
produced the alloy now known as D9. Finally, two groups of alloys were
examined in detail, one consisting of ferritic alloys and the other of high-
nickel superalloys. Both gave favourable results, although the superalloy

precipitate phases (γ') were altered by irradiation so that the mechanical properties changed. Some indications of the variations of swelling with composition in the Fe-Cr-Ni system are given in Figs. 4.12 and 4.13.

There is no good theoretical basis at present for guiding the selection of alloys for low swelling characteristics. One phenomenon that may well play an important role in this and other aspects of in-reactor behaviour is called radiation-induced segregation (RIS) [19]. It has been shown that vacancies drag along with them solutes with a certain size relationships to the host lattice; generally a little less in size than the host. Hence silicon is dragged along in stainless steel until the vacancy comes to rest at a void or at a surface of another kind. This can have several consequences: on the one hand the pile-up of solute around a void prevents its growth, so there may be a basis for limiting swelling. On the other hand *depletion* of the surfaces in chromium with a corresponding build-up of smaller atoms has been observed. This could reduce corrosion-resistance at the very place where it is most needed.

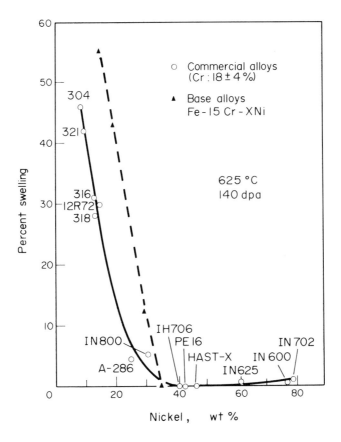

Fig. 4.12. Swelling of Fe-Cr-Ni alloys as a function of composition. <u>Source</u>: B. R. T. Frost.

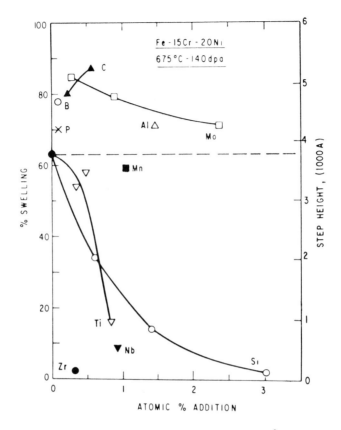

Fig. 4.13. Swelling of Fe–Cr–Ni alloy at 675°C and 140 dpa
as a function of various minor alloy additions.

What is very clear is that the behaviour of the minor alloying elements in
alloys can drastically influence radiation effects and that we know relatively
little about the fundamentals of these processes.

From a designer's viewpoint there is a lot of data on clad swelling and on
the effect of irradiation on mechanical properties. Attempts have been made
to cast the data in the form of equations which can be used over a wide
range of design conditions. Figure 4.14 shows the summarized data for
annealed and cold worked Types 304 and 316 stainless steels together with
swelling equations. The available data is documented in the *Nuclear Systems
Materials Handbook* [20]. However, there is a problem of generating radiation
effects data in a timely manner. Even if a large breeder reactor like FFTF
is available for conducting irradiation tests it takes on the order of three
years or more to generate fluences representative of core lifetimes in
commercial reactors. The swelling versus fluence curves for most alloys are
nonlinear in character, so that extrapolations are not valid. There is
typically an "induction period" up to $\sim 5 \times 10^{22}$ n/cm^2/sec before swelling
starts, after which it follows a reasonably consistent pattern. The exact
duration of the induction period cannot be predicted as yet, and it can vary
from alloy to alloy.

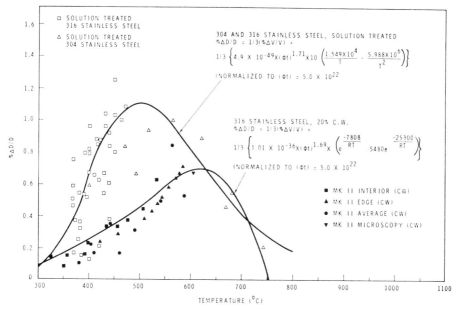

Fig. 4.14.   Swelling curves and equations for Types 304
and 316 stainless steel. Source: Hanford
Engineering Development Laboratory.

The only alternative to length reactor irradiation is ion simulation [16].
If stainless steel is bombarded with 4 MeV Ni$^+$ ions, the ions come to rest in
a narrow band $\sim$1 μm from the surface, creating considerable damage within
that band (Fig. 4.15). In principle one can simulate accelerated fast neutron
effects by such a means. To be realistic one must inject helium to the same
depth as the nickel at the same time. The amount of void swelling may be
measured by interferometry, and the void number and size distribution may be
viewed by transmission electron microscopy. Indeed, the main justification
for the building of high voltage ($\sim$1 MeV) electron microscopes has been the
fact that one can produce electron damage and view it simultaneously; it will
soon be possible to study ion damage in the same way. Ion simulation speeds
up the radiation damage rate by $\sim$10^3 times, so that instead of a reactor
irradiation of several years, we may produce the same damage in a few hours.
As one might expect, the kinetics of the void formation process are somewhat
altered, causing the swelling versus temperature curve to be shifted $\sim$100°C
along the temperature axis. However, the method is very useful for comparing
alloys or for screening a group of alloys; it should not be used for design
data except to help in throwing light on very high dose effects, e.g. to
determine whether swelling ever saturates with dose.

To permit a comparison of ion and neutron damage doses, the concept of
"displacements per atom" or dpa was introduced. One can calculate the
average number of times each atom will be displaced from its lattice site in
a given damage down, be it neutrons or ions, so that a comparison can be made:
10^{23} neutrons cm^{-2} dose corresponds to about 60 dpa.

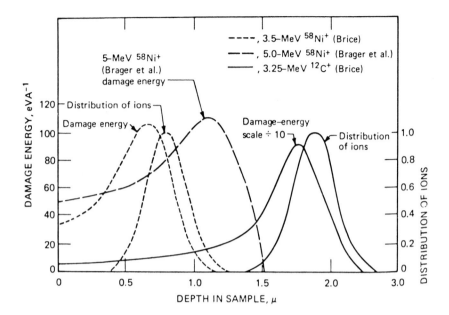

Fig. 4.15.   Damage energy, and distributions of stopped ions normalized to 1, for 3.5- and 5.0-MeV $^{58}Ni^+$ and for 3.25-MeV $^{12}C^+$ on a stainless-steel target.   <u>Source</u>: A. Taylor, ANL.

REFERENCES

1. Greenfield, P., *et al.*, Geneva Conference A/CONF 28/P. 146 (May 1964).
2. Lustman, B. and Kerze, F., *Metallurgy of Zirconium*, McGraw-Hill, New York (first published in 1955).
3. Douglass, D. L., *The Metallurgy of Zirconium*, IAEA Review (Suppl.) (1971).
4. Cox, B., Oxidation of zirconium and its alloys, *Advances in Corrosion Science*, 5, M. Fontana and R. W. Staehle (eds.), Plenum Press (1976).
5. Pickman, D. O., *Physical Metallurgy of Reactor Fuel Elements*, The Metals Society, pp. 437-438 (1975).
6. Ambortsumyan, R. S., *et al.*, *Proceedings of Second International Conference on Peaceful Uses of Atomic Energy*, Geneva, Vol. 5, p. 12 (1958).
7. *MATPRO - a Handbook of Materials Properties*, P. E. MacDonald and L. B. Thompson (eds.), ANCR-1263: NRC-5 (Feb. 1976) and NUREG/CR-0497: TREE-1280 Rev. 1 (Feb. 1980). Obtainable from NTIS, Springfield, VA, and the US Nuclear Regulatory Commission.
8. INCO publishes data sheets on its products. INCO, World Trade Center, New York, NY.
9. Walter, C. M. Golden, G. H. and Olson, N. J., U-Pu-Zr Metal Alloy: A Potential Fuel for LMFBR's, Report ANL-76-28 (November 1975). Available from NTIS or TIC at Oak Ridge, TN.
10. IAEA Panel Proceedings, 1974: Behavior and Chemical State of Irradiated Ceramic Fuels.

11. Snyder, R. B., Natesan, K. and Kassner, T. F., *J. Nucl. Mat.* <u>50</u>, 259 (1974).
12. Billington, D. S. and Crawford, J. S., *Radiation Damage in Solids*, Princeton University Press, NJ (1961).
13. Thompson, M. W., *Defects and Radiation Damage*, Cambridge University Press, England (1969).
14. Olander, D. S., *Fundamental Aspects of Nuclear Reactor Fuel Elements*, TID-26711, TIC-USERDA (1976), chapters 17, 18 and 19.
15. Cawthorne, C. and Fulton, E., *Nature*, <u>216</u>, 576 (1967).
16. Corbett, J. W. and Ianniello, L. C. (eds.), *Radiation-Induced Voids in Metals*, AEC Symposium Series No. 26, USAEC (1972).
17. Huebotter, P. R. and Bump, T. R., ibid, p. 84.
18. Frost, B. R. T., Radiation damage considerations, in *Fundamental Aspects of Structural Alloy Design*, R. Jaffee and B. Wilcox (eds.), Plenum Press, p. 521 (1977).
19. Workshop on Solute Segregation and Phase Stability during Irradiation, reported in *J. Nucl. Mat.* <u>83</u> (1) (1979).
20. *Nuclear Systems Materials Handbook*, compiled by Hanford Engineering Development Laboratory for the US DOE. In general, available only to DOE contractors.

CHAPTER 5

Fuel Element Design and Modelling

The design, development, testing and licensing of a fuel element is a lengthy undertaking performed by numerous people whose efforts are often coordinated by committees. The need for a new fuel element design usually originates in an idea for a new design of reactor, but it can also come from the desire to upgrade the performance of existing reactor cores. Today, after 35 years of reactor development, it is highly probable that one will build on existing knowledge to design a new fuel element rather than attempt to develop a radically new concept, especially in view of the licensing problems that are likely to arise in the latter case.

The starting point of a fuel element design is when someone or some body of people decide on the broad design specification for a particular type of reactor that is to meet a given need. That need might be for a power reactor of a given size and performance, for a materials testing reactor or to meet some political need such as national defence or the prevention of the proliferation of nuclear weapons. In any event, a reactor concept is drawn up specifying the thermal power output, the coolant characteristics and other features of its performance. An example is the Gas-cooled Fast Reactor concept developed by the General Atomics Corporation (GA), which had previously developed the High-temperature Gas-cooled Reactor (HTGR). GA's idea was to use the high-pressure helium coolant technology and the prestressed concrete vessel concept of the HTGR in conjunction with a fast reactor core that drew heavily from the experience of the Liquid Metal-cooled Fast Breeder Reactor (LMFBR) (Fig. 5.1). A specification was drawn up for a 360 MWe, 1090 MWt demonstration reactor with a helium coolant pressure of 1520 psi, circulated at high velocity (Table 5.1). Along with such a specification came the normal requirements of a breeder power reactor core: high burn-up (>10 a/o) without cladding failure, ease of fuel handling, ability to survive moderate power and flow transients, etc.

TABLE 5.1.   Specification for Gas-cooled Fast Reactor

Design data		
Overall plant	Thermal power, MWt	1090
	Net electric power, MWe	360
	Net plant efficiency, %	33
	Net plant heat rate, Btu/kWh	10,200
	Number of main loops	3
Reactor	Fuel material	Pu/U oxide
	Cladding material	austenitic stainless steel
	Fuel rod O.D., inch	0.315
	Fuel rod pitch, inch	0.453
	Fuel rod spacing	grids
	Fuel rods/assembly	265
	Fuel assemblies	150
	Control and shutdown assemblies	19
	Blanket assemblies	162
	Shield reflector assemblies	138
	Active core height, inch	47.2
	Linear power rating, peak/avg., kW/ft	11.3/6.3
	Reactor inlet temperature, °F	568
	Reactor outlet temperature, °F	975
	Helium pressure, psi	1520
	Helium flow rate, lb/sec	2050
Power conversion system	Helium loop pressure drop, psi	34
	Circulator power, each, BHP	14,300
	Steam generator inlet helium temperature, °F	968
	Superheater outlet temperature, °F	905
	Superheater outlet pressure, psi	1550
	Feedwater temperature, °F	340
	Turbine cycle	straight expansion
	Turbine speed, rpm	3600
	Turbine inlet pressure, psi	1450
	Turbine inlet temperature, °F	900
	Steam flow rate, total, 10^6 lb/hr	3.49

Courtesy: General Atomic Co.

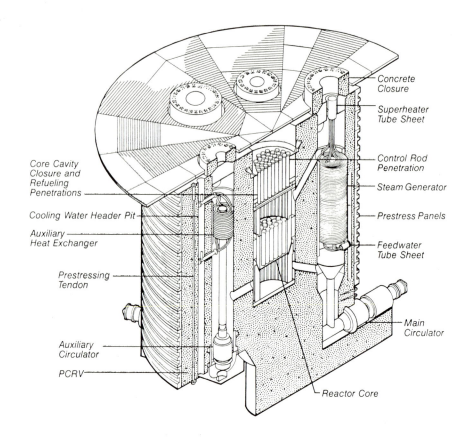

Fig. 5.1.   GCFR nuclear steam supply system.
Source: General Atomic Company.

The first response to a reactor design specification is to perform reactor
physics calculations to determine the possible ranges of fuel enrichment and
the distribution and ratios of the fuel, cladding, coolant, moderator and
control materials throughout the core.   These calculations will also yield the
neutron flux and spectrum as a function of core position, so giving the heat
output per unit volume or power density (H) as a function of position.

$$H = N\sigma_f \phi E \tag{5.1}$$

where N = fissile atom density per unit volume,
      σ_f = fission cross-section,
      ϕ = neutron flux,
      E = energy per fission event (~200 MeV).*

---

*2.696 x 10^{21} fissions produce 1 Mwdt.

Due to neutron leakage, ϕ will decrease away from the core centre and hence the heat generation rate will decrease with increasing distance from the core centre.

The next step is to perform heat transfer calculations to determine the feasible core size and geometry from the standpoint of heat removal, recognizing that the answer will be influenced by the purpose of the reactor. Hence, for a research reactor one will probably wish to maximise the flux ϕ, requiring a high power density, while for a thermal power reactor a much lower power density may be desirable.

A basic equation for the rate of heat removal from a reactor core is:

$$q = (\rho \cdot c_p) \cdot v \cdot A_f \cdot \Delta t_c \qquad (5.2)$$

where $\quad q$ = heat removed per second, i.e. rate of heat removal,
$(\rho \cdot c_p)$ = volumetric heat capacity of the coolant,
$\quad v$ = coolant velocity in ft/sec or m/sec,
$\quad A_f$ = coolant contact area in the core,
$\quad \Delta t_c$ = temperature change from the bottom to top of the core.

Gaseous coolants have low values of $(\rho \cdot c_p)$ even at high pressures so that a high gas velocity and a high value of A_f is needed. The latter can be aided through extending the surface area by fins which also promote turbulence and increase the heat transfer rate. With a coolant like liquid sodium, with a high heat capacity and a very low film drop from cladding to the bulk coolant, extended surfaces and high flow velocity are not necessary.

There are several thermal limits on core design which will influence fuel and cladding dimensions: the coolant must not boil (or at least depart from nucleate boiling), the fuel must not melt at its centre, the fuel-cladding interface temperature must be below that for significant reaction or alloying over the fuel element lifetime, and the cladding must remain sufficiently cool to retain adequate mechanical strength. These thermal limits can be calculated from a knowledge of the heat generation and removal rates and the heat flow path from the bulk fuel to the bulk coolant.

In performing core temperature calculations one has to be aware of the sum of the uncertainties or local variations that each parameter may experience. To take a simple example, cladding diameters are specified to be within a size range such as 0.4 in $\pm$ 0.002 in which allows a variation of 0.004 in in cladding diameter. The sum of all such variations of the same, positive sign gives a "hot channel factor" or worst case which is assumed in order to provide some safety margin.

To return to equation (5.2), there is a basic limitation on flow velocity: high flow rates require a high pumping power and also may induce vibrations in the subassemblies. The pumping power (p) per unit volume flow (per gallon or litre) is given by:

$$p = \frac{f \cdot \rho \cdot L \cdot v^2}{2D} \qquad (5.3)$$

where f = friction factor (related to roughness),
$\quad \rho$ = coolant density,
$\quad L$ = coolant channel length,
$\quad D$ = coolant channel diameter,
$\quad v$ = coolant velocity.

Note that the pumping power increases as the square of the coolant velocity.

Another factor that determines core geometry is the boiling behaviour of the coolant, if it is a liquid. Water undergoes a transition from nucleate boiling to film boiling, or departure from nucleate boiling (DNB), with "dry-out" and overheating of the cladding, at a heat flux somewhat in excess of 10^6 Btu/hr/ft^2/$^\circ$F or $\sim$100 cal/sec/cm^2/$^\circ$C. This is a real limitation in the design of water-cooled-and-moderated cores for research reactors where high heat fluxes are required. The much higher boiling point and different boiling characteristics of sodium mean that similar limitations are absent, which explains why sodium is a good coolant for high power density fast reactor cores.

Next it is necessary to determine the fuel diameter, before one can determine channel diameters (rod spacing) and fuel element length. This brings us to the difficult question of the choice of fuel type. The limits of performance are determined by different phenomena in the different fuels. In metallic fuels the fundamental limitation is the reaction kinetics, i.e. the time-temperature relationships, for fuel-cladding reactions. This arises because reactor physics and economics may forbid the use of chemically inert or non-reactive cladding materials, such as molybdenum. Hence one uses a commercially available alloy with an acceptable neutron capture cross-section. For example, the U-Fs fuel of the EBR-II reactor is clad in Fe-Cr-Ni stainless steel. U-Fe and U-Ni eutectics (liquid phase) form at 725°C and 740°C, respectively, hence one must avoid the fuel-cladding interface getting into this temperature range for long enough for appreciable reaction to occur. It used to be argued that fuel swelling, which increases with temperature, was the most serious limitation on performance. However, the EBR-II Mk II fuel element design [1] overcame that problem by incorporating a sufficiently large sodium bond to allow the fuel to swell $\sim$25% before it touched the cladding (Fig. 3.7). At that stage the swelling drops off drastically because the bubbles interconnect and release their gas to the plenum.

Oxide fuels, on the other hand, are very inert in contact with stainless steel or Zircaloy and their limitation lies in their low thermal conductivity, which varies with temperature. The temperature difference from fuel surface to centre of a cylinder of fuel is:

$$\Delta T_{cs} = \frac{\sigma_f \cdot \phi \cdot N \cdot r^2}{4k} \cdot E = \frac{H \cdot r^2}{4k} \qquad (5.4)$$

where σ_f = fission cross-section
ϕ = neutron flux (locally) $\left.\right\}$ H is defined in eq. (5.1)
N = fissile atom density
r = fuel radius
k = fuel thermal conductivity
E = energy per fission

Note that k varies with temperature in UO$_2$ (see Fig. 2.11) so that it is usual to think in terms of $\int_{T_1}^{T_2} k \cdot d\theta$ rather than a single value of k. The generally accepted limit on UO$_2$ or (U, Pu)O$_2$ performance is centre melting ($\sim$2800°C), largely because the consequences of fuel melting are not well understood. Thus, one should be able to determine the maximum value of r if one knows the fuel surface temperature; if we set $\Delta T_{cs} = T_{melt} - T_{surface}$, all parameters in eq. (5.4) are then known except r. However, the determination of the fuel surface temperature is uncertain because of the nature of the fuel-cladding interface. This consists of a fairly smooth but lightly oxidized stainless steel surface in contact with a rough ceramic, the interstices being filled initially with helium, at pressures varying from 15 to 1000 psi, and later during fission in the core are diluted with the fission gases xenon and krypton. One attempt [2] to define the heat transfer coefficient h_2 derived the equation:

$$h_2 = \frac{k_f}{0.5\ C_{op} + C\ (R_1 + R_2) + (g_1 + g_2)} + \frac{km \cdot p}{H\ (R_1{}^2 + R_2{}^2)^{1/4}} \qquad (5.5)$$

where R_1 and R_2 are the two surface roughnesses,
    k_f = thermal conductivity of the filling gas,
    km = thermal conductivity of the solid,
    g_1 and g_2 are accommodation coefficients,
    p = interfacial pressure,
    C_{op} = average operating clearance,
      C and H are constants.

The first term on the right-hand side of the equation describes the heat transfer through the gas phase while the second term describes the heat transfer across the solid-solid interface. Typically a fuel surface will be $\sim$300-400°C above the bulk coolant temperature. Furthermore, all designs using oxide fuel stay well away from the fuel melting temperature — generally operating at $\sim$2000°C centre temperature, although the CANDU and AGR designs stay below $\sim$1500°C to keep fission gas release to a minimum because they do not have gas plenums.

To recapitulate, knowledge of the fuel surface temperature, the fuel thermal conductivity and the fission rate allows us to define an upper bound on rod diameter set by the fuel melting temperature.

Carbides and nitrides are better conductors than oxide. Combined with their high melting temperatures they could be designed to have much larger diameters.

Dispersion fuels (UAl_3 in Al and UC_2 in graphite, for example) generally have thermal conductivities close to that of the matrix, which is one reason for their existence, i.e. to run at temperatures close to that of the coolant. The plate-type fuels for research reactors have a high surface area to volume ratio, a good matrix thermal conductivity and good restraint on swelling of the fuel particles at modest temperatures.

Returning to metallic and oxide fuels, a further consideration is the cladding thickness. In sodium-bonded metal fuels this can be low because of the absence of stresses. For oxides in thermal reactors the neutron capture in the cladding sets an upper limit on its thickness. In fast reactors it is a combination of neutron capture and thermal stress that limits the thickness. The thermal stress σ_t is given by:

$$\sigma_t = \frac{\alpha \cdot E}{(1 - \nu)K} \qquad (5.6)$$

where α = coefficient of thermal expansion,
    K = thermal conductivity,
    E = Young's modulus,
    ν = Poisson's ratio.

For stainless steel α is large and K is small, so that a smaller thickness could be tolerated than, say, for the refractory metals molybdenum, niobium or vanadium. The lower limit to cladding thickness is set by stress calculations with a "wastage allowance" added to account for fuel-cladding and coolant-cladding interactions.

Thus, through considerations of materials properties and heat transfer one can arrive at a compromise value of the fuel rod diameter and cladding thickness, having in the process selected the best fuel and cladding combination from a thermal standpoint. One also has to consider a myriad of other factors, including fuel and cladding irradiation behaviour.

Having made what one believes to be the best choice of fuel and cladding to suit the application and having decided on the best compromise (optimized) dimensions of the fuel pin one must proceed to test these choices.

The reader is reminded of Table 1.1 in the first chapter, which is a "road-map" to fuel element development. We have just discussed stages 1 through 6 in Table 1.1. We will now consider steps 7 through 10. Step 7, fabrication procedures, will not be elaborated on here since they are discussed in other chapters. One need only note that the first stages of fabrication development are on a small (0.1 to 1 kg) scale, where flexibility is more important than economics. Scale-up to production does not occur until step 12, after a considerable amount of testing and analysis has been carried out.

Step 8 in our scheme of fuel element design and development is modelling. Fuel element models are mathematical descriptions of the various complex phenomena that occur when a fuel element is in a reactor core. Models are usually time-dependent in nature, i.e. they analyse the phenomena as a function of reactor power. There are two basic types of model, the first being of a fairly simple nature in which a thermal analysis is combined with an elementary mechanical analysis that incorporates the various sources of stress to make a prediction of the cladding state and hence the probability of failure. The SWELL code for oxide [3] and the BEMOD code for metal fuels [4], both developed at Argonne, are examples of this type of code.

The second type of model attempts to treat fuel element behaviour rigorously, modelling all the phenomena in the most realistic way and using finite element or other sophisticated mechanical analyses to predict the lifetime of the fuel element. LIFE [5], CYGRO [6], and COMETHE [7] are examples of this type of code.

Fuel element *models* are phenomenological descriptions of materials behaviour. The *codes* are the translations of these models into computer language. Modelling is the art of constructing models, placing them in computer language (codes), inserting the necessary materials property values and computing the likely behaviour of a fuel element under certain conditions. One might look on modelling as the theoretical description of fuel element behaviour. To anticipate a later discussion, the models are tested and refined by using them to plan irradiation experiments in which they predict the outcome. After the experiments are concluded the experimental parameters are compared with the predictions and the models are examined for sources of disagreement. This is an iterative process in which one improves the process of planning irradiation experiments as well as refines and improves the models. The predictive capability of a model is highly dependent on the ability to describe important phenomena, such as fuel swelling and gas release, and on the availability of valid property data for the fuel and cladding. Indeed, models are valuable in assessing the relative importance of the input data and the need for measurements, i.e. for scoping fuel and cladding property measurement programmes. There is a class of radiation experiments, known as "separate effects tests", which focus on measuring specific properties or phenomena, such as irradiation creep or fission gas release. These are valuable inputs to models.

When one has developed or obtained a model or code in which one has confidence, it can be used to reduce the amount of expensive in-pile testing that is needed. It may be used to interpolate between, or extrapolate from, a small number of experimental data points. It is in this sense that codes are often used by regulatory agencies in the licensing of new cores. Figures 5.2 and 5.3 illustrate the complexity of phenomena that occur in the fuel and cladding of an operating fuel element. Figure 5.2 separates the effects in the fuel and the cladding in terms that were discussed in the preceding chapters. Figure 5.3 brings the fuel and cladding together and pinpoints fuel-cladding interactions. However, it does not illustrate how these interactions vary with time and how they lead to fuel element failure. A fuel element model attempts to do this in a rigorous manner.

We will illustrate the simple and the complex models respectively by means of the SWELL and the LIFE codes for describing the performance of fast reactor fuel elements made up of $(U,Pu)O_2$ fuel and stainless steel cladding. First, the sequence of events in the life of such a fuel element will be described in phenomenological terms so that we can see what the model must be capable of describing.

Typically a fast reactor fuel element is made up of 0.23 in diameter $(U,Pu)O_2$ sintered pellets of 90% theoretical density (10% porosity) fitted tightly inside a Type 316 stainless steel tube with a 0.015 in wall thickness (Fig. 5.4). The average gap between fuel and cladding is 2-3 mils and the filling atmosphere is of helium. The fuel column (core height) is 36 in (90 cm) with a UO_2 axial blanket 18 in long at either end, and a top plenum 36 in long. The pin is sealed by a weld and is held in a subassembly of $\sim$200 pins, which are spaced by wire wraps or by grids inside a hexagonal duct made of stainless steel measuring $\sim$5 in across the flats and 0.10 in in thickness (Fig. 5.5).

The subassembly is loaded into the core and is cooled by upward flowing sodium that enters at $\sim$400°C and exits at 600°C when the reactor is at full power. After core loading is complete the absorber rods are withdrawn or fuel control rods inserted and the core power steadily increases until the design power is reached. During that time fission occurs at a steadily increasing rate in the fuel, producing heat, fission products and neutrons. The neutrons pass through the cladding and move around the core. When the reactor is at full power the fuel is fissioning at a rate that produces $\sim$30-40 kW/m of heat for each fuel pin, i.e. the 221 pin subassembly is generating $\sim$6 MW of heat. This heat rating establishes a large thermal gradient from the fuel centre to the surface, on the order of 2000°C over 0.115 in (0.275 cm), i.e. a gradient of $\sim$7000°C/cm. This has several consequences. First the thermal stresses induced in the brittle part of the fuel cause it to crack, usually radially to form pie-shaped segments (Fig. 5.6 oxide elements). Above $\sim$1500°C the fuel is plastic and can accommodate the stresses. Second, significant forces act on the sinter pores and cause them to move up the temperature gradient. Near the centre of the pellet the movement is due to vaporization of fuel from the hotter side of the pores and condensation on the colder side, but surface diffusion processes may also play a role. This pore migration occurs above $\sim$1800°C, and results in the formation of a central hole. The pore migration also forms

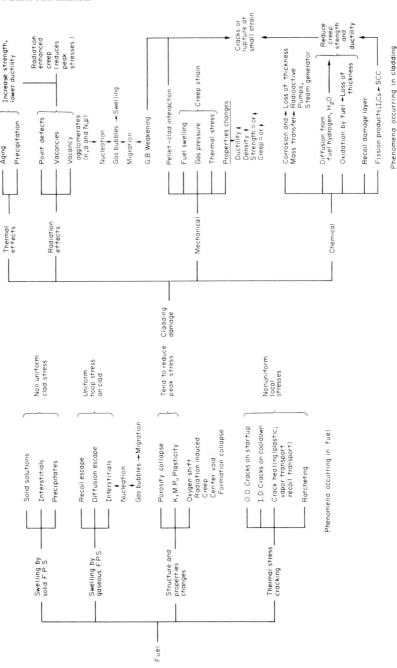

Fig. 5.2.   Phenomena occurring in cladding and fuel.
Source: Zebroski and Levenson, Ann. Rev. of
Energy 1, 101 (1976).

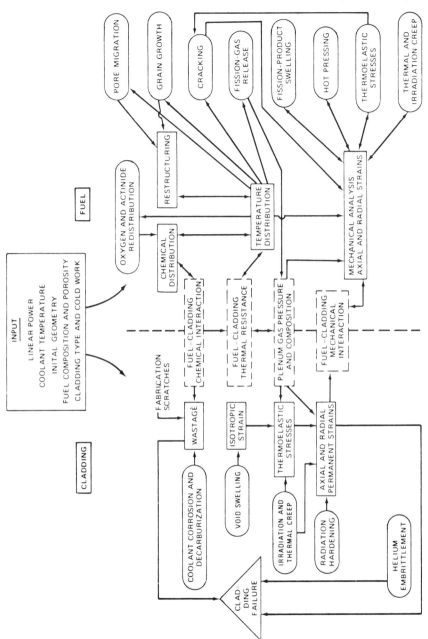

Fig. 5.3. Interrelation of mechanical, metallurgical,
and chemical processes in fuel-element
irradiation behavior. Source: D. Olander,
Fundamental Aspects of Nuclear Reactor Fuel
Elements, TID-26711.

columnar grains from the central hole out to the ~1800°C isotherm (Fig. 5.7). Between 1800° and 1500°C equiaxed grain growth occurs. Below 1500°C the fuel structure changes very little, but this region is initially cracked. As radiation proceeds these cracks may heal.

A minor effect in breeder reactors, but of major importance in thermal reactors, is the disappearance of the small sinter pores (<1 μ diameter) during the first few hours of irradiation. This densification has significant effects in shrinking the 12-ft fuel column of an LWR fuel element, but is of little consequence to the 3-ft LMFBR fuel column. This is further discussed in a later chapter on LWR fuel elements. Another densification process is hot pressing; that is, the fuel swelling and fission gas pressures on the fuel cause the larger pores to disappear through a conventional hot pressing process. This seems to be a more important densification process than fuel creep; at one time it was thought that the cladding restraint caused the fuel to creep and swell in on itself, i.e. to use the central void to accommodate swelling, but the evidence is against this.

Following restructuring the fuel is now denser and is in firm contact with the cladding. At steady power, continued fission gives a steadily increasing inventory of fission products, which we noted earlier could be divided into several classes:

1. Insoluble gases: Xe and Kr
2. Volatiles: I, Cs
3. Soluble in (U,Pu)O : Zr, Ce, Nd
4. Insoluble oxides: Ba, Sr
5. Insoluble metals: Mo, Tc, Ru, Rh

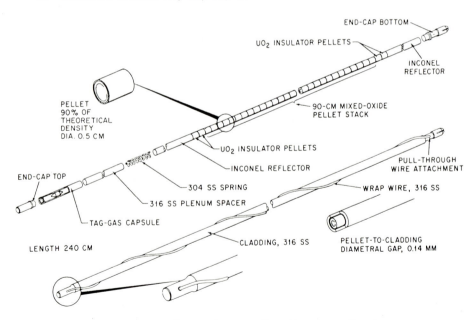

Fig. 5.4.   Fuel pin of the Fast Test Reactor.   Courtesy: C. Burgess, Hanford Engineering Development Laboratory.

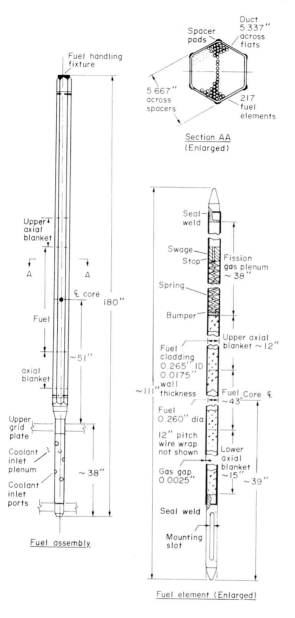

Fig. 5.5.   LMFBR subassembly.   Source: ERDA-1535.

Helium – bonded mixed oxide
at 14 kW/ft

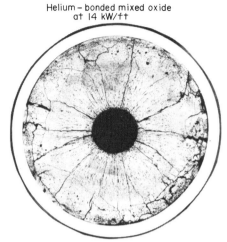

ANL Element SOPC-3 at 3 at.%
Center temp:~2400° C

Helium –bonded mixed oxide
at 25 kW/ft

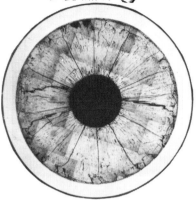

ANL Element SOV-7 at 3.6 at.%
Center temp: 2800° C

Helium –bonded mixed carbide
at 25 kW/ft

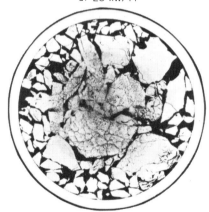

ANL Element HMV-4 at 10.6 at.%
Center temp:~1650° C

Sodium – bonded mixed carbide
at 25 kW/ft

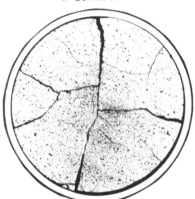

Ward element W8X at 6.5 at.%
Center temp:~1100° C

Microstructural comparison
of 0.290-in. diameter elements
of 80-85% smear density
irradiated in the EBR-II

Fig. 5.6.   Microstructural comparison of 0.290 in diameter
elements of 80-85% smear density irradiated in
the EBR-II.   Source: ANL.

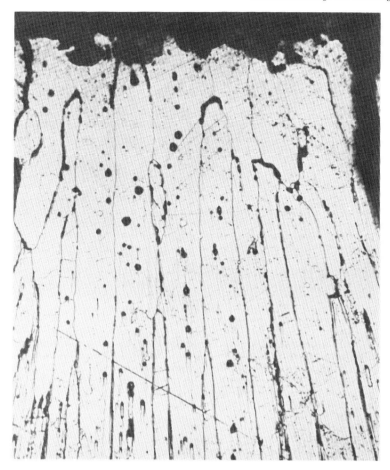

Fig. 5.7.   Bubble trails in columnar grains of
$(Pu_{0.2}U_{0.8})O_2$ fuel.   Source: F. A. Nichols,
ANL.

The gaseous fission products are insoluble in the fuel and form bubbles that migrate and coalesce, giving rise to fuel swelling, and they migrate from the bulk of the fuel to the central void and the fuel-clad gap and the plenum. The rates of swelling and gas release are very important in determining fuel element behaviour.

Extensive data are available on swelling and gas release. Figure 5.8 shows data on the swelling rates of mixed oxide fuel irradiated at two different centre temperatures to high burn-ups. Figure 5.9 shows fission gas release data from mixed oxide fuel pins at various centre temperatures to high burn-ups. From such data and from detailed postirradiation studies of bubble sizes and fission product distribution, models of swelling and gas release have been developed. These two processes combine to exert stresses on the cladding.

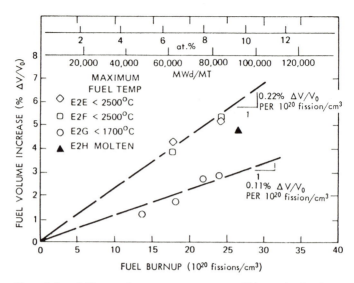

Fig. 5.8.   Effect of temperature on swelling of mixed-
oxide fuel.   Source: R. Duncan *et al.*, Fast
Reactor Fuel Element Technology, ANS, 1971.

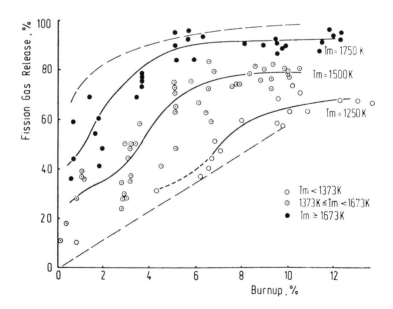

Fig. 5.9.   Fission gas release in mixed oxide fuel pins
as a function of burn-up.   Source: H. Zimmermann,
KfK, Karlsruhe, Germany.

The volatile elements may combine to form caesium iodide, but the experimental evidence suggests that they migrate down the temperature gradient to the fuel-clad gap and along that gap to the coolest regions. The main practical consequence of this is that chemical reactions with the cladding may occur that weaken the clad wall. In LWRs the iodine probably induces stress-corrosion cracking in the Zircaloy cladding after it has migrated to the fuel-clad gap. In LMFBRs molybdenum forms uranium molybdate which swells compared to the fuel.

As irradiation proceeds the cladding will be subjected to uniform and steadily rising stresses due to fission gas pressure and to nonuniform stresses as fuel swelling forces the pellets or cracked portions of pellets into stronger contact with the cladding. At the pellet-pellet interfaces and across thermal stress cracks discontinuities in stresses on the cladding will occur. As the reactor power is changed the interfacial pressures will change. Obviously a mild overpower transient, e.g. due to control rod movement, will cause greater cladding strain. A rapid shutdown or start-up will cause rapid changes in cladding strain (and in the stress pattern and crack patterns in the fuel pellets).

We should turn our attention to what is happening to the cladding. Remember that the cladding is our first line of defence and a primary objective of modelling is to build up an inventory of what is happening to the cladding that might cause it to fail. Chapter 4 described the phenomena that occur during the irradiation of stainless steel by fast neutrons. In the fast reactor fuel element, neutron irradiation causes the cladding first to harden and to lose ductility through defect cluster formation, then to swell through void formation and also to creep at a more rapid rate through the formation of vacancies. The time-dependent interactions of the fuel and the cladding are complex because the clad swelling tends to move the cladding away from the fuel and to reduce fuel-induced stresses, while the radiation-induced creep may cause further distention of the cladding under the action of the fission gas pressure. One thing is certain — the amount of deformation that the cladding can undergo before failure is very small (less than 1%). Fracture tends to be intergranular in nature as the grain interiors are hardened and nonuniform ductility is very low. We are, therefore, very interested in the stress-strain relationships in cladding alloys as a function of radiation dose (fluence).

Let us now see how the SWELL and LIFE codes handle these phenomena and how successful they are in predicting experimental results. SWELL was developed as a fast-running subroutine in a core design code. It is basically a thermal and mechanical analysis. The fuel element is divided up as shown in Fig. 5.10. The fuel is divided into ten or more "axial nodes", i.e. it is chopped up into ten or more equal cylinders along its length. Then thermal, nuclear and mechanical analyses are performed on each axial node and the boundary conditions are matched at the nodel interfaces. The fuel is considered to restructure at time zero and to touch the cladding. This greatly simplifies the thermal analysis. Then, for a given time step, the amount of fission gas release and fuel swelling is calculated (see Fig. 5.11). The plenum pressure is calculated from the amount of gas there, the plenum volume and plenum temperature, assuming that the perfect gas laws hold.

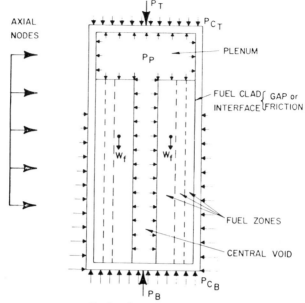

Fuel-element Axial Cross Section

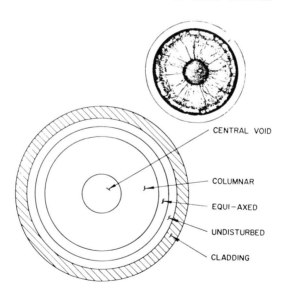

Fuel-element Radial Cross Section

Fig. 5.10.   Fuel-element axial cross section; fuel-element
radial cross section. Source: ANL.

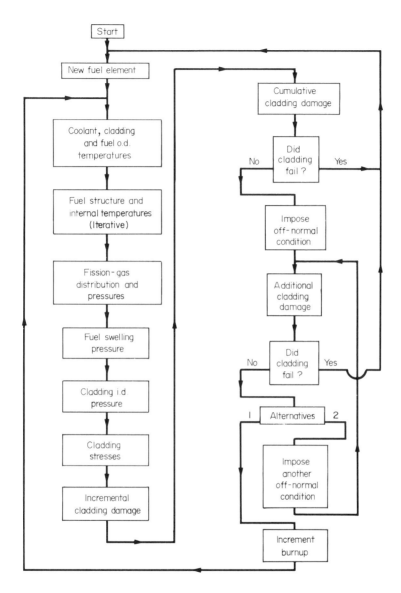

Fig. 5.11.   SWELL flow diagram.   Source: ANL.

The swelling pressure (P_{FS}) on the cladding is calculated to be

$$P_{FS} = f_1 \ (P_{R1}) + f_2 \ (P_p) \tag{5.7}$$

Where f_1 is an adjustable function fitted to give agreement with clad strain data and f_2 treats the fuel and cladding as two elastic hollow cylinders in contact with P_p (the plenum pressure) acting on the fuel cylinder inner diameter. P_{R1} is the fuel swelling pressure at node 1. This is a deliberately over-simplified procedure using adjustable parameters to force a fit to benchmark data. At any axial position the sum of the two pressures, P_{FS} and P_p, is considered to be the effective pressure on the cladding.

SWELL does not follow the fuel power history in any detailed way, only recognizing full power and zero power operation. However, the code does attempt to handle creep and fatigue behaviour of the cladding by summing the creep (primary and secondary) and fatigue due to power cycling. At the conclusion of each time interval in the calculations, all damages which have occurred up to that time are summed at each axial position and if any of the sums is equal to or greater than unity, cladding failure is assumed to occur at that axial position.

At the end of each time interval the code checks whether any of several possible off-normal conditions could cause cladding failure. Off normal conditions cover sudden changes in power and coolant flow. The effects of these on cladding damage are assumed to be only in the primary creep behaviour. Incremental local primary creep strains are calculated and added to the accumulated damage up to that point in time. Again, if the sum exceeds unity failure is predicted. The code assumes that the fuel element experiences only one off-normal event during its lifetime, generally near the end of its normal life.

For a "quick and dirty" prediction of fuel element lifetime SWELL was fairly successful. However, fuel element developers, designers, reactors operators and regulatory bodies required a more rigorous approach which followed the power history and did not oversimplify the complex processes occurring during the life of an operating fuel element. LIFE was one of several responses to that need. These codes are rigorous analyses based on first principles and are intended to follow the power history.

LIFE started as a code to describe mixed oxide-stainless steel fuel elements in a fast reactor core, but has been expanded to include carbide and nitride fast reactor fuels and oxide LWR fuels. The LMFBR oxide fuel version is now used in its fourth version LIFE-4 and is still being improved (Table 5.2). Here we will discuss LIFE-II, because it is well documented, and will comment on later modifications. It is worth noting at this point that LIFE and other fuel element codes are on file at:

>National Energy Software Center
>Applied Mathematics Division, Building 223
>Argonne National Laboratory
>9700 S. Cass Avenue
>Argonne, Illinois 60439, USA

The code can be obtained on tape and IBM cards with a user's manual.

One important difference between SWELL and LIFE is the latter's ability to follow the reactor operating history in detail. The code uses an axisymmetric, generalized plane strain mechanical analysis with an iterative procedure for each time step. "Axisymmetric" means that there is no tangential variation in any of the important parameters. "Plane strain" implies that although the fuel and cladding may move axially (not necessarily at the same rate), planes perpendicular to the Z (vertical) direction in each material remain plane during deformation.

TABLE 5.2.   Evolution of the LIFE Code

LIFE-I (Jankus and Weeks)

A.   1-D (Quasi 2-D) steady state, thermomechanical analysis of fast reactor fuel elements (mixed oxide, helium bond, stainless steel cladding).

B.   Three fuel rings (columnar zone, equiaxed zone, unrestructured zone), and one cladding ring.

C.   Empirical laws to establish fuel-ring boundaries.

LIFE-II (Jankus, Weeks, Poeppel)

A.   Additions

   1.   Fuel cracking and healing model and power cycling capability.
   2.   Cladding failure model based on stress rupture and time fraction.

B.   Modifications

   1.   Fuel and cladding properties.
   2.   Recalibration.

LIFE-III (Poeppel, Billone and others [9].   Users' Manual

A.   Additions

   1.   Multi-ring fuel and cladding analysis (up to 20 rings).
   2.   Continuous porosity migration and grain growth models (PORMIG).
   3.   Nonequilibrium fuel swelling model (treated like hot pressing).
   4.   Anisotropic fuel swelling model.
   5.   New gap conductance model (GAPCON) [10].
   6.   Cladding primary creep model.
   7.   Cladding wastage model.
   8.   Strain fraction option to failure model.
   9.   Large strain structural analysis for fuel and cladding.
  10.   Central-hole closure model (switch boundary conditions for closed central hole).

B.   Modifications

   0.   Correction of serious error in stress analysis (S4 modification).
   1.   Fuel properties (fuel creep).
   2.   Cladding properties (creep, swelling, and time to rupture).
   3.   Recalibration and check out to a larger set of fuel pin irradiations.

C.   Improvements

   1.   Option to input cladding o.d. temperatures.
   2.   Option to change constants in QQ subroutine (materials constants).
   3.   Improved output — more geared to designer's use.

TABLE 5.2.   (Continued)

Spin-offs from LIFE-III

A.   Advanced Fuels

    1.   Carbide and nitride fuel properties and models.
    2.   Helium and sodium bond properties.
    3.   Stainless steel and 11 advanced alloy properties.
    4.   Transient heat transfer.
    5.   Calibration.

       UNCLE-T
       Carbide and nitride
       transient behaviour

B.   GCFR

    1.   Vented pin options.
    2.   Smooth-ribbed cladding option.

C.   Light Water Reactor

    1.   Radial flux depression.
    2.   Fuel densification model (sintering).
    3.   Zircaloy axial growth model.
    4.   Water coolant properties and models.

D.   Research Reactors (LIFE-PLATE)

LIFE-4

A.   Transient Heat Transfer and TREAT Capsule Options.

B.   Fuel Primary Creep.

C.   Fuel Melting (volume expansion upon melting).

D.   PFRAS Option for Transient Gas Release and Swelling.

E.   Oxygen Migration Model.

F.   Cladding Plasticity Model (time-independent).

G.   Reorganization of Common Blocks and Code Structure to Minimize Storage
    Requirements and to Optimize Running Time.

H.   Fuel and Cladding Properties.

    1.   Advanced alloy properties.

Spin-offs from LIFE-4

A.   Advanced Fuels (LIFE-4CN)

B.   Metal Fuels (LIFE-METAL)

The fuel element in LIFE-II is represented by the geometry shown in Fig. 5.10. The fuel is generally divided into four radial zones — undisturbed, equiaxed grain growth and columnar grain growth, plus a central void although a larger number of radial zones is allowed. The fuel element is divided into ten or more axial sections. An initial fuel-cladding gap is assumed and the gap is allowed to open and close any number of times depending on the reactor power conditions.

The flow chart for LIFE-II is shown in Fig. 5.12. As a function of time and for each axial position the code predicts the temperature distribution; the fuel and cladding dimensions, including length changes; the extent of fuel cracking and crack healing; the extent of fuel restructuring; the amount of gas released to the plenum and the plenum pressure; the migration and distribution of the actinides, oxygen and the fission products; hot pressing of the fuel; the stress and strain distribution for each region of fuel and cladding; the fuel-cladding interfacial pressure; and the amount of cumulative damage that has occurred in the cladding. This last parameter can be used to state the probability of cladding failure by comparison with experimental data. Unfortunately, as with a number of input and output parameters in this and all other codes, inadequate experimental data exist to allow full use of these codes. Thus, for example, one can make a meaningful comparison between the predicted and measured cladding profile or the fission gas released to the plenum, but one cannot compare the predicted and actual time to cladding failure. One can, perhaps, compare the code's prediction of the probability of failure with a plot of fuel element failures versus burn-up, rod power, etc.

As the flow chart indicates, once the operating conditions for a given time step have been determined the temperature distribution is calculated for each axial section, factoring in the fuel-cladding interfacial pressure calculated in the previous time step. Then the boundary migration rates for each of the three fuel regions is established; this only changes significantly during the early stages of the fuel element life.

The amount of fission products generated in the time step are calculated, knowing the fission rate during that time period. The pressure of the plenum can then be calculated.

Then the code selects an axial section and performs a deformation analysis on it. An initial approximation is made for the change in total strains in the fuel and cladding that are expected to occur in each region during the time step, based on the length of the time step and the changes that occurred in the previous step. From this assumption, first approximations of the average deviatoric and hydrostatic stresses in each concentric cylinder are determined from empirical creep laws and the swelling and hot pressing relationships in the fuel. These approximations are then used in a complex analysis to derive a second approximation for the change in total strains in each region and this is compared with the initial approximation. If convergence criteria on the stress and strain are not met, this iterative process continues until they are. The code has built into it maximum allowable changes of stress, strain and restructuring for a given time step. If these are exceeded, the code cuts the length of the time step and recalculates. It will also cut the time step if convergence is too slow. If the power has changed from the previous time step, the local thermal stresses in each fuel region are computed and a determination is made of the amount of fuel cracking. The mechanical properties of the fuel are adjusted according to the calculated amount of fuel cracking, crack-healing and restructuring, so that the effects on the cladding may be calculated.

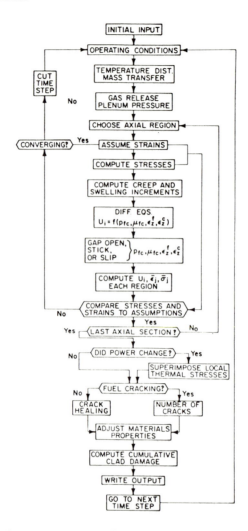

Fig. 5.12.   Simplified LIFE-II flow chart.
Source: Jankus and Weeks, ANL.

The main purpose of this explanation is to describe the conceptual process rather than to give the details of the use of the code. The reader is referred to other literature [5,8] and to the Argonne National Energy Software Center for such details. We will, however, demonstrate the predictive capability of the LIFE code with some examples and finally comment on recent modifications to the code.

The test of a code is obviously how well it predicts the results of an inreactor test. In fact, code development uses this process to find the weak spots in the code and to improve it.

LIFE-I had all of the basic ingredients described above, but it did not allow for fuel cracking, thus the outer regions of the fuel acted as a "pressure vessel" to contain the hotter fuel, giving low values of the cladding stress. LIFE-II was developed to incorporate fuel cracking. This version was checked out against a fuel element designated X072-SOPC5 which had been irradiated in the EBR-II reactor and subjected to a full post-irradiation examination (Table 5.3 gives details). The model calculations were run in four ways; the full operating history was run with and without fuel cracking and the "equivalent" steady-stage average operating power history was run with and without fuel cracking. Figure 5.13 shows the restructuring analysis and Fig. 5.14 shows the cladding hoop stress. The calculations using fuel cracking and following the reactor power fluctuations come closest to predicting the observed fission gas release and fuel element diameter change. It is important to note that the "steady-state" calculations yielded lower values of cladding hoop stress than the load-following calculations. As one would expect, the closer the calculations represent the real-life conditions, the better is the agreement with experimental results. Indeed, this is the impetus for code refinement and improvement.

TABLE 5.3.   LIFE-II Results for Fuel Element X072-SOPC5

	Experimental values	Full history (274 days) with fuel cracking	without fuel cracking	Steady operating power (275 days) (8.1 kW/ft) with fuel cracking	without fuel cracking
1. Total fluence, NVT	$\sim 2.2 \times 10^{22}$	2.21×10^{22}	2.21×10^{22}	2.24×10^{22}	2.24×10^{22}
2. Clad swelling strain, %	~ 0.13	0.14	0.14	0.16	0.16
3. Clad creep strain, %	–	< 0.001	0.21	< 0.001	< 0.001
4. $\Delta D/D$ final, %	~ 0.17	0.17	0.37	0.18	0.19
5. Time of gap[b] closure, hr	gap closed	850	125	~ 1620	~ 3200
6. Outer radius of columnar zone, in	~ 0.100	0.106	0.095	0.043	0.043
7. Fission gas released, %	56.4	56.2	55.3	50.8	51.4

Fuel: Mixed-oxide pellets (20 wt % Pu); Clad: 304H SS; Burn-up: 3.0 at %.

[a]Results shown are for section just below midplane.
[b]SOPC5 had an initial diametral fuel-clad gap of 14 mils.

Source: Weeks and Jankus, ANL.

Obviously, different fuel types have different physical characteristics that lead to different versions of the codes. In addition, different reactor types call for different codes or code versions. Versions of the LIFE code (called UNCLE) have been developed for LMFBR carbide and nitride fuels which, as we have seen above, have higher thermal conductivities than oxides and run at lower centre temperatures and lower thermal gradients. Hence more fission gas is retained in the fuel so that the UNCLE-S code has a different fuel swelling model from LIFE. These fuels crack differently from oxide; and a different cracking model is used.

Metal fuels are generally used in sodium bonded fuel elements in which the bond is sized to allow enough swelling to reach the breakaway regime, i.e. at ~ 25–30% swelling. The BEMOD code was developed to handle this fuel element concept, although LIFE has also been adapted to this metal fuel design.

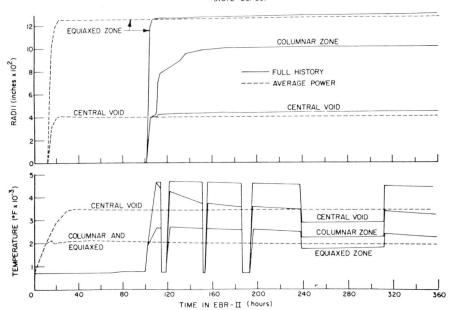

Fig. 5.13. Zone boundaries and temperatures for element
X072-SOPC5, run both steady-state and with
full operating history. Source: Jankus and
Weeks, ANL.

In light-water reactors the fuel is still oxide: UO_2 rather than $(U,Pu)O_2$,
but their properties are similar. However, the cladding is zircaloy and the
external pressure is high (1000 or 2000 psi for the BWR and PWR, respectively).
The modelling of commercial LWR fuel elements has lagged behind those for
LMFBRs and for naval reactors. Recently the Electric Power Research Institute
sponsored a comparison of codes for commercial LWRs and selected the Belgian
Comethe code for general use. Figures 5.15 and 5.16 show a comparison [11]
of the capabilities four codes in predicting the fuel central temperature
and the fission gas release, respectively. Table 5.4 shows predictions of the
Comethe-II code for Saxton fuel elements.

GRASS-SST calculates the birth of inert gases upon fission, their atomic
migration, formation of bubble nuclei, bubble migration, coalescence and
re-solution, the effects of obstacles and the behaviour of bubbles on grain
boundaries. The subroutine calculates fission-gas-induced swelling and
fission gas release as a function of time for steady-state and transient
conditions. A system of coupled equations for the evolution of gas bubble
distributions is developed, having the general form:

$$F_i^\alpha = - a_i^\alpha F_i^\alpha - b_i^\alpha F_i^{\alpha 2} + C_i^\alpha \tag{5.8}$$

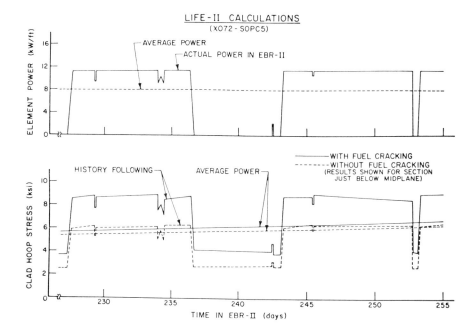

Fig. 5.14.   Element power and cladding hoop stress for
element X072-SOPC5, run both steady-state
and with full operating history. Source:
Jankus and Weeks, ANL.

where i has values from 1 to N, $\alpha = 1$, 2, 3. F_i^α represents the number of
α-type bubbles in the ith size class per unit volume. The equation represents
the rate at which α-type bubbles are lost from (grow out of) the ith size
class due to coalescence with bubbles in that class; b_i^α represents the rate
at which α-type bubbles are lost from the ith size class due to coalescence
with bubbles in *other* size classes, etc., while c_i^α represents the rate at
which bubbles are being added to the ith class from other classes.
Figure 5.18 is a flow chart for one annular fuel region out of the many into
which the fuel has been divided. From these calculations the swelling and
gas release can be calculated. Figure 5.19 compares calculated and measured
gas release for fuel from the H. B. Robinson, Saxton and CVTR fuel.

GRASS-SST was developed especially to permit the computation of bubble
behaviour during reactor overpower transients, either of a mild or a severe
nature. The results of these computations are incorporated in the FRAP
code [16] that calculates fuel element behaviour in a transient situation
much as LIFE or COMETHE do for normal operating conditions. There is
evidence for increased bubble mobility during transients [17], possibly due
to an increased rate of atom attachment to and detachment from the bubble
surfaces, or due to increased dislocation sweeping of bubbles. Also bubbles
may become overpressurized during transient heating because vacancies cannot
flow in fast enough to restore equilibrium. If the overpressure results in
an equivalent stress in excess of the yield stress, local plastic deformation
will occur, increasing the density of dislocations around the bubbles. The
intersection of the bubbles and the dislocations produce ledges which

facilitate atom attachment and detachment. Thus one calculates the time T_i^y for the bubble overpressure to generate a stress equivalent to the UO_2 yield stress and the time T_i^B, the bubble relaxation time to equilibrium. Thus as T_i^y decreases and T_i^B increases the system departs more from equilibrium. One can then proceed to calculate the enhanced gas release in a transient of a given magnitude. Comparisons have been made between such calculations and fuel tested in the Power Burst Facility at INEL, Idaho and in Direct Electrical Heating (DEH) simulation experiments [17]. Figure 5.20 shows this comparison for DEH-heated PWR fuel from H. B. Robinson with PFB tested fuel from the Saxton reactor. Qualitatively the structures are similar.

TABLE 5.4. COMETHE-II Prediction of Saxton Fuel Performance

Comparison with Pins Irradiation in Saxton

Pellet diameter: 1.376 cm   Bonding gas assumed to
Initial radial gap: 0.15 mm   be helium
                                                  Sheath: stainless steel
Pin No. 1
    Bulk density: 97% TD
    Maximum linear power: 760 W/cm
    Irradiation history simulated by 6 successive levels

Item	Experiment	Prediction
Burn-up [MWd/MT (HM[a])]	11,620	10,500
Central temperature (°C)	2340	2460
Gas release (cm^3 STP)	73.6	98.9

Pin No. 2
    Bulk density: 97% TD
    Irradiation history simulated by 3 successive power
      levels
    End of irradiation after a sharp power peak at
      568 W/cm

Item	Experi- ment	Prediction Before peak	End of peak
Burn-up [MWd/MT (HM[a])]	4500	4220	
Central temperature (°C)	1925	1938	2191
Gas release (cm^3 STP)	15.81	15.1	

Pin No. 3
    Density 78% TD (in the central part of the pin)
    Maximum linear power: 592 W/cm
    Irradiation history simulated by 5 successive power
      levels

Item	Experiment	Prediction
Burn-up [MWd/MT (HM[a])]	10,000	9660
Central temperature (°C)	2200	2241
Gas release (cm^3 STP)	82.4	70.0

[a]HM refers to heavy metal.

Source: R. Godesar *et al.*, *Nucl. Appl. & Tech.* **9**(2), 205 (1970).

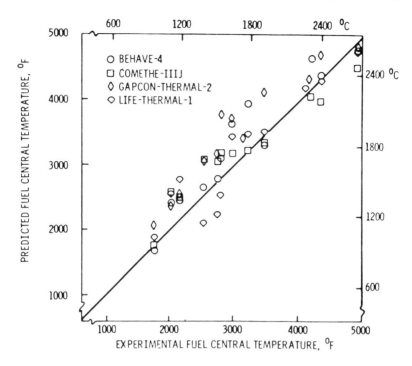

Fig. 5.15. Phase III fuel temperatures. <u>Source</u>:
M. G. Andrews *et al.*, Water Reactor Fuel
Performance, ANS, 1977.

Before we leave the subject of fuel element modelling it must be re-emphasized
that developing the model is only half of the job. The other half is the
measurement of the physical and chemical properties that must be inserted into
the many subroutines of the codes. The results are only as good as the input
data. Oddly enough, while there is still a lot of effort being put into
modelling and code development, the efforts to generate input data have
dropped off alarmingly in the late 1970s. Data exist for most important
properties of most of the fuels. However, the quality of that data is very
variable and much of it needs improvements in precision, accuracy and range
of measurement. Mention was made in earlier chapters of the MATPRO compila-
tion of LWR fuel and cladding data (Chapter 4, ref. 7) and the Nuclear
Systems Materials Handbook for LMFBR data. These compilations are in forms
that are easily compatible with fuel element codes.

The primary aims of fuel element modelling are to provide a scientific basis
for fuel element design and to help in planning fuel element tests and in
the analysis of these tests (steps 9 and 10 in Table 1.1). Some practical
aspects of these tests are discussed in the next two chapters. Once a design
has been tested and refined (step 11), the fuel element passes into the
commercialization stage (steps 12, 13 and 14). Chapters 8 through 11 describe
the design and performance of commercial or near-commercial fuel elements.

In closing this chapter, the reader is reminded that it is the intention to describe here the methodology of fuel element design and development in broad principles. The practice of fuel element design requires years of exposure to the application of detailed engineering, physics and materials science principles with an overlay of economics and the exercise of many compromises to reach a workable design, something that can only be inferred in a book.

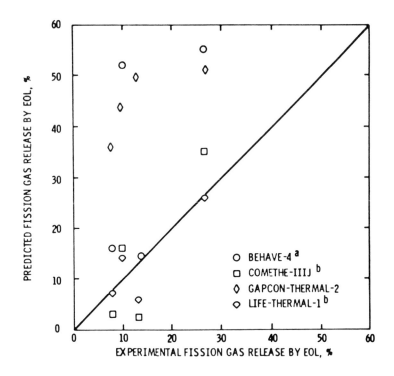

a Phase III gas release model was used.
b Phase III Zircaloy-creep model was used.

Fig. 5.16.   Phase III fission gas release.   Source:
M. G. Andrews *et al.*, Water Reactor Fuel
Performance, ANS, 1977.

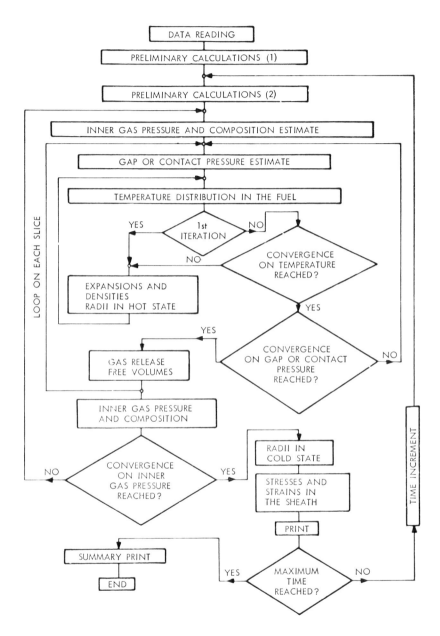

Fig. 5.17.   Comethe Code flow chart.   Source: Godesar
et al., Nucl. Appl. & Tech. 9, August 1970.

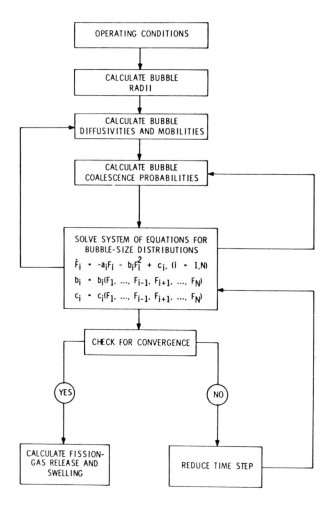

Fig. 5.18.   Grass-SST flow chart.   Source: J. Rest, ANL.

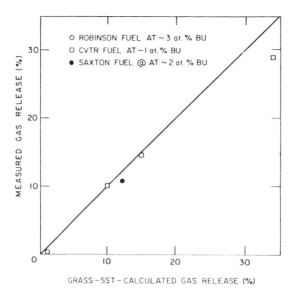

Fig. 5.19. Comparison of GRASS-SST predictions with end-
life gas release. Source: J. Rest, ANL.

Comparison of posttest fractographs

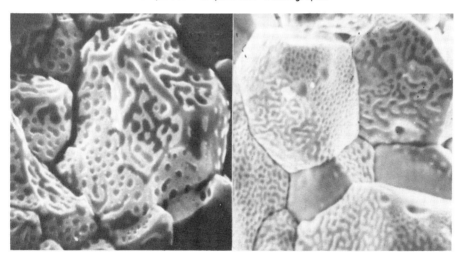

DEH-tested robinson   ⊢——————⊣   PBF-tested saxton
                              2 μm

Fig. 5.20. SEM fractographs of Robinson and Saxton fuels;
Saxton specimen. Source: S. Gehl, ANL.

SOME SOURCES OF DATA AND CODES

Codes

National Energy Software Center
Applied Mathematics Division, Building 223
Argonne National Laboratory
9700 S. Cass Avenue
Argonne, Illinois 60439, USA

Fuels and Cladding

Nuclear Systems Materials Handbook
Hanford Engineering Development Laboratory
P.O. Box 1970
Richland, Washington 99352, USA

MATPRO - A Handbook of Materials Properties
for Use in the Analysis of LWR Fuel Rod Behavior
ANCR/1263 USNRC   NRC-5, Pub. 1976
Available from NTIS

MATPRO - Version II (Revision 1)
NUREG/CR-0497
TREE-1280, Rev. 1, Pub. 1980
Available from NTIS

Fuels

Uranium Ceramics Data Manual
DEG Report 120 (R)
UKAEA, HMSO, Great Britain

Physical Properties of Some Plutonium Ceramic Compounds
TRG Report 1601 (R)
HMSO, Great Britain

Properties for LMFBR Safety Analysis
ANL-CEN-RSD-76-1
Argonne National Laboratory

Steels

Steels Data Manual
TRG Report 840 (R)
UKAEA, HMSO, Great Britain, 1965

REFERENCES

1. Walter, C. M., Olson, N. J. and Hofman, G. L., *Proceedings of the AIME Symposium on Materials Performance in Operating Reactors*, p. 181 (1973).
2. Garnier, J. E. and Begej, S., PNL-2696 (1979) PNL-3232 (1980).
3. Bump, T. R., "SWELL: A Fortran II Code for Estimating the Lifetimes of Mixed-Oxide Fuel Elements", ANL-7681 (1973).
4. Jankus, V. Z., "BEMOD — a Code for the Lifetime of Metallic Elements", ANL-7586 (1969).
5. Jankus, V. Z. and Weeks, R. W., *Nucl. Eng. Des.* 18, 83–96 (1972).
6. Freidrich, C. M. and Gullinger, W. H., WAPD-TM-547 (1966).
7. Godesar, R., Guyette, M. and Hoppe, N., *Nucl. Appl. & Tech.* 9 (2), 205 (1970).
8. Olander, D., "Fundamental Aspects of Nuclear Reactor Fuel Elements", TID-26711, p. 566 (1976).
9. Billone, M. C., *et al.*, "LIFE-III Fuel Element Performance Code", ERDA 77-56 (1977).

10. Hann, C. R., Beyer, C. E. and Parchen, J. L., BNWL-1778 (1973).
11. Andrews, M. G., Freeburn, H. R. and Pati, S. R., *Proceedings of a Topical Meeting on Water Reactor Fuel Performance*, American Nuclear Society, St. Charles, IL, p. 104 (1977).
12. Gittus, J. and Howl, D., ibid, p. 169.
13. Notley, M. J. F., *Nucl. Appl. & Des.* 9, 195 (1970).
14. Poeppel, R. B., *Proceedings of a Conference on Fast Reactor Fuel Element Technology*, American Nuclear Society, p. 311 (1971).
15. Rest, J., GRASS-SST, NUREG/CR-0202, ANL-78-53 (1978).
16. Dearien, J. A., *et al.*, "FRAP-53 Computer Code," TFBP-TR-164 (1978) and see NUREG/CR-0786 (1979).
17. Rest, J. and Gehl, S., *Nucl. Eng. & Design*, 56 (1), 233–256 (1980).

CHAPTER 6

Fuel Element Performance Testing and Qualification

Having arrived at a conceptual design for a fuel element and a subassembly, it must be tested and refined. The actual sequence depends on whether one is starting from scratch with a new reactor concept or merely developing an improvement to an existing concept. In the former case it will probably be necessary to undertake fairly extensive development of fuel and fuel element fabrication techniques, followed by fuel and cladding property measurements out-of-pile and in-pile, then small-scale fuel element testing on through single subassembly testing to partial and full core loadings. The scale of the fabrication development will match the scale of the testing. Thus in the early stages of development the fabrication process will be a "bench-top" or $\sim$100 g batch scale which is adequate to provide samples for property measurements and small-scale irradiation tests. However, great care must be taken to ensure that there is continuity in the fuel composition and micro-structure at all stages of development, i.e. the early samples and measure-ments should be applicable to production fuel and cladding. This can only be achieved through rigorous quality assurance procedures.

In this chapter we will address the kind of testing that is done; in the next chapter the hardware will be described. The stages of fuel element testing are in effect stages 7 through 14 of Table 1.1 with emphasis on step 9. We will describe them in a little more detail and with a different notation:

1. Fabricate fuel samples, e.g. vacuum melt and cast a metallic alloy; make powder, press it and sinter it into pellets for an oxide ceramic; arc melt and case rods of a refractory metal or metal-like compound.
2. Characterize the samples by chemical analysis, optical metallography and thermal analysis. This includes measurement of the fuel composition, grain size and microstructure. In the case of ceramic fuels the porosity is described in detail and the stoichiometry — or ratio of metal to nonmetal atoms — is carefully measured.
3. Measure physical properties such as thermal conductivity, strength and ductility, and diffusional properties to high temperatures.
4. Study the compatibility (reaction kinetics) of the fuel with candidate cladding materials and coolants. Modify the fuel composition, if necessary, to improve compatibility.

5. Irradiate small samples in a test reactor to determine swelling and gas release characteristics. These may be clad or unclad samples, with or without in-reactor sensors.
6. Fabricate prototypic fuel elements (usually of the final radial dimensions, but shorter in length than the intended element) and irradiate in a test reactor, possibly in a loop or in an experimental subassembly. In the latter case the element may be encapsulated to protect the reactor core against unexpected failure modes.
7. Examine these irradiated pins, which may have been modelled — if so a comparison between predicted and actual performance is made.
8. Fabricate a full size subassembly. This may be unfuelled for hydraulic testing, i.e. placing in a coolant loop for pressure drop and heat-transfer measurements. A fuelled version will be placed in a loop in a test reactor or in a power reactor core. This subassembly and the single elements will have to have their enrichment level adjusted to give the correct linear power in the test reactor flux and spectrum. We will address this aspect further in the next chapter.
9. Perform interim and final post-irradiation examination on the bundle elements. Reserve some elements for transient testing. Give feedback on performance to the designer, the modeller, the fabricator and the person who measures properties so that adjustments and improvements to the design can be made.
10. Place fabrication on *at least* pilot production scale (∿1 to 3 kg per batch) and make a partial or full core loading of a new reactor. Only at this stage can one begin to collect statistical data on performance, including the all-important data on failure occurrence rate as a function of burn-up or neutron exposure.
11. As a result of large-scale, statistical testing plus post-irradiation analysis, devise and incorporate modifications to the design.
12. Obtain a licence from the Nuclear Regulatory Commission or other appropriate national body to operate this new core.

Some types of reactor (probably most) will require some transient testing. Generally this is at a scale of not more than one subassembly and often on a smaller scale. For fast reactors in the USA the TREAT reactor is available for overpower transient testing (which should be upgraded in the early 1980s to allow it to take longer fuel elements). The CABRI reactor at Cadarache in France has a similar purpose. The SLSF* loops are available in the Engineering Test Reactor for flow blockage and pin-to-pin failure propagation tests. A Safety Test Reactor has been designed to test groups of sub-assemblies, but funding has not yet been approved.

For LWRs the Power Burst Facility (PBF) can test single subassemblies for overpower transients while the LOFT is a *loss-of-flow-t*est. Both are at INEL in Idaho, USA.

Because the later stages in the testing sequence are very costly and lengthy, modelling and simulation testing are important aids to generating greater confidence in a design and to permitting the extrapolation and interpolation of the meagre experimental data.

An example of a simulation technique is the Direct Electrical Heating experiment [1] to reproduce transient conditions fairly inexpensively and quickly. A column of fuel pellets is held between two metal electrodes and

---

*Sodium Loop Safety Facility.

heated by ohmic heating, i.e. low voltage, high current dc electricity is passed through the fuel column (Fig. 6.1). The ohmic heating reproduces fission heating fairly well, especially in UO_2 whose electrical conductivity increases with rising temperature. The fuel column may be unclad, representing a late stage in a core accident, or clad in transparent silica or even in metallic cladding. Temperatures are measured (generally by an optical pyrometer) and a high-speed cine film is taken of the sequence of events as the fuel is taken above its melting point.

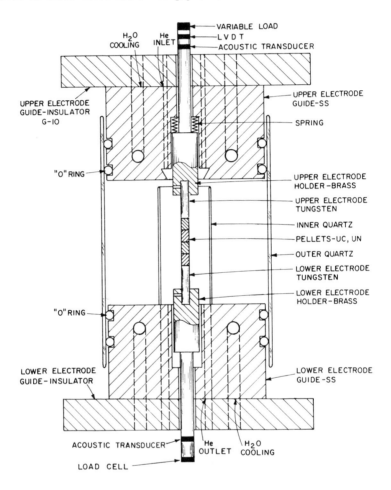

Fig. 6.1.   Schematic of specimen holder for DEH rig.
Source: B. Wrona *et al.*, Nuclear Technology, 20, November 1973.

The fuel may be pre-irradiated, i.e. extracted from a spent fuel element, in which case one can measure the rate of release of fission gases and volatile fission products in a gas sweep system and a multichannel analyser. Experiments have been performed on oxide, carbide and metal fuels with results that compare well with transient reactor tests and with model predictions.

A recent example of fuel development and testing is that for an advanced proliferation-resistant fuel for research reactors. Since many countries that do not possess nuclear weapons operate research reactors fuelled by fully enriched uranium, a programme is under way to reduce this enrichment without sacrificing reactor performance. One candidate fuel is U_3Si, because it has a high uranium atom density compared to UAl_3 or U_3O_8: thus one should get the same core performance with a lower level of enrichment. U_3Si was developed in Canada by workers [2] as a bulk fuel for power reactors, but it had some problems of corrosion with water. For research reactors it will be dispersed in an aluminium matrix and rolled into sheets.

The development programme for this fuel has involved first searching out all the available information from Canada and elsewhere. Next small samples were made by arc melting to examine both the crushing characteristics (to make powder particles) and corrosion behaviour in water. It was decided to add aluminium to improve the corrosion resistance, which led to some ternary phase diagram studies. In parallel with these studies, tungsten carbide particles were used as a stand-in for U_3Si in studying its dispersion in aluminium and incorporation into a "picture-frame" type fuel element (see Fig. 1.3). Then samples were fabricated with naturally enriched U_3Si for laboratory and engineering tests [3]. Small plates with appropriately enriched uranium were then made for irradiation in the core of the Oak Ridge Research Reactor (ORR). Assemblies of plates for flow testing are probably not needed, because the exact geometry of exiting research reactor fuel assemblies will be reproduced. Provided the ORR tests are successful (i.e. reach the design burn-up without failure or distortion) production elements for insertion in a representative research reactor core will be made and tested. Finally, commercial vendors will be sought to produce these elements for the world market.

STRATEGIES

Until a commercial reactor is built and operated, fuel performance testing for a particular reactor type has tended to be a "bootstrapping" operation. That is, the best use is made of the available test facilities to proceed in a series of logical steps until one has sufficient confidence to design and build a commercial-sized reactor. The statistics and confidence level in proceeding along this path are generally not as good as they should be.

The fast breeder reactor programme in the USA (and in other countries) may be used to illustrate this point. In 1951 Walter Zinn brought the EBR-I fast reactor to power, following a minimum of fuel element testing. By present standards the fuel elements were small and were not expected to achieve high burn-ups. However, the reactor worked well for several years and allowed the testing of more advanced fuel element designs, leading to the EBR-II reactor. This reactor was designed around pyrometallurgical reprocessing (see Chapter 10), but it evolved into an irradiation test reactor. The core was still small — only 13.5 in high and 21 in diameter — but it could accept experimental fuel pins of varying types and diameters. The types of experimental subassembly are described in the next chapter. A typical core loading is shown in Fig. 6.2.

Key: SSTH—SST thimble   P   —½ driver fuel, ½ SST
BETH—Beryllium thimble   B   —Depleted uranium
SSCR—SST control rod   R   —SST reflector
C # —Control rod   MKII —Mark II fuel
S # —Safety rod   HW —High worth control rod
D —Driver fuel   ISD —INST. S/A dummy
                        SD —Structural dummy

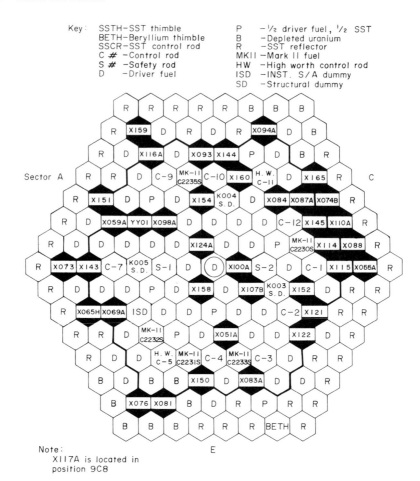

Note:
X117A is located in
position 9C8

Fig. 6.2.   EBR-II Core loading for reactor run 58A. The
numbered solid hexes are experimental subassemblies and the "D" type driver fuel subassemblies contain mark IA elements.

Irradiations of mixed oxide pins in the EBR-II core gave performance data on
which to base the FFTF driver core design. However, the FFTF is 36 in high
(see Fig. 5.7), so that length effects were tested in the GETR reactor,
which is a thermal test reactor. Hence the experimental pins were of low
uranium enrichment and were irradiated in pairs, rather than bundles, to
avoid self-screening or flux-depression effects.* EBR-II differs from FFTF
and commercial reactors in one very important respect other than core length:
the neutron flux is significantly lower. This has two consequences; first,
the fuel irradiated in EBR-II must contain highly enriched uranium in order

---

*Absorption of thermal neutrons by uranium atoms in the outer fuel layers
reduces the flux to the inner fissile atoms so that the fission rate and
burn-up reduces from surface to centre of a fuel rod.

to achieve an appropriate fission rate. Thus, more fissions occur in uranium than in plutonium, giving rise to a different fission product yield and to different chemical effects. Second, for a given burn-up in EBR-II, the cladding fluence is almost an order of magnitude lower than it would be in a power reactor. Such fluence : burn-up ratio effects are difficult to extrapolate or to model. Hence there is some uncertainty in using the EBR-II results for performance prediction.

Transient effects were studied by exposing pins that had been previously irradiated in EBR-II, to appropriate overpower and loss of flow tests in the TREAT reactor and in SLSF loops in the ETR reactor. TREAT is being modified to accept FFTF pins for testing.

FFTF will permit both "standard" testing of mixed oxide subassemblies of commercial size and of other fuel types in special loops. All of this will precede the construction of a power reactor. FFTF is too valuable to be used for risky experiments such as operation with a number of failed elements, yet this condition would be desirable in a commercial reactor. With FFTF on line for fuel testing to failure, EBR-II will be used to study operation beyond cladding breach (RBCB), since a forced shutdown of EBR-II would no longer jeopardize the whole breeder programme.

Mention was made of the bootstrapping approach to obtaining incremental improvements in performance. Walter, Olson and Hofman [4] discussed an alternative approach, using Weibull statistics. The Weibull equation is:

$$F(t) = 1 - \exp\left[-\frac{t - t_0}{\eta}\right]^{\beta} \qquad (6.1)$$

where $F(t)$ = the cumulative probability of failure,
$\quad\quad \beta$ = Weibull slope (a constant evaluated from failure data),
$\quad\quad \eta$ = characteristic life, $t = \eta + t_0$, at which 63% of the failures
$\quad\quad\quad$ will occur,
$\quad\quad t$ = response variable (time, burn-up, etc.),
$\quad\quad t_0$ = origin of the distribution.

If this is applied to EBR-II driver fuel the equation for the Mk II fuel becomes:

$$F(BU_{max}) = 1 - \exp\left[-\left(\frac{BU_{max}}{9.9}\right)^{15.1}\right] \qquad (6.2)$$

where $F(BU_{max})$ = cumulative cladding failure probability as a function of
$\quad\quad\quad\quad$ burn-up,
$\quad\quad (BU_{max})$ = maximum burn-up.

Application of this equation to the case where there shall be less than a 5% chance of having 1 cladding failure in 10,000 pins, gave a burn-up limit of 4.7 at % of heavy atoms.

The method may be used for the statistical design of irradiation experiments as follows:

1. Establish which parameter limits operation.
2. Determine the range of environmental variables (pin power, flux, fission rate, etc.) and design variables (fuel composition, cladding properties, etc.) present in the reactor system of interest.
3. Design experiments to study the limiting parameter through this range of variables or for the most severe combination of these variables.
4. Determine the fraction of elements that exceed the limit (i.e. fail) as a function of burn-up, etc.
5. Perform a metallurgical evaluation (post-irradiation examination).
6. Perform a Weibull analysis.

To anticipate a later chapter, the Nuclear Regulatory Commission (NRC) requires for each reactor plant (or new core type) that testing, inspection and fuel surveillance plans be submitted and reviewed [5]. The plans then become part of the plant's safety analysis report. The NRC publishes an annual report [6] which summarizes the fuel performance data for all the commercial fuels in operating reactors, including the number of failures and their causes.

When fuel is being tested (or simply irradiated) in power reactors the operators (and the fuel vendors) look generally for evidence of failed fuel elements by monitoring the coolant for fission products and sometimes for delayed neutrons. In addition, an attempt is made to "sip" each fuel assembly. That is, samples of coolant are removed during shutdown at the outlet of the assembly and analysed for fission products to warn of failures or to identify failed assemblies.

In the discussion of fuel element modelling we mentioned "separate effects tests", which are designed to measure a specific effect as a function of design variables. Typical of such tests is the measurement of fuel swelling as a function of burn-up. Figure 6.3 shows the results of irradiation tests on a large number of plate-type UO_2 fuel elements irradiated in pressurized water over a range of fuel thickness, density and temperature [7]. Surprisingly the results are well grouped and define two curves — an initial rate of 0.4%/1 at % burn-up where the swelling is partially accommodated by the fabrication porosity, and an "intrinsic" swelling rate of 1.7%/1 at % burn-up. Such data are valuable for modelling studies and for fuel design where one can, for example, adjust the initial porosity to shift the transition to intrinsic swelling to a higher burn-up level.

POST-TEST EXAMINATION

While a few in-reactor tests may use *in-situ* instrumentation, this is often unreliable and tests must generally be analysed nondestructively and then destructively. Nondestructive examination (NDE) may be performed either in a storage pool at the reactor (a "poolside examination"), which is common for commercial fuels, or in a hot cell. Obviously destructive examination has to be carried out in a hot cell because of the risk of spreading fission products and other radioactive material.

NDE may include profilometry (clad diameter versus length over several angular sectors), gamma scanning for certain isotopes (e.g. ^{137}Cs and $^{95}Zr-^{95}Nb$), straightness, cladding integrity (presence of cracks) and X-ray, gamma ray or neutron radiography. This can be performed at poolside or in a hot cell.

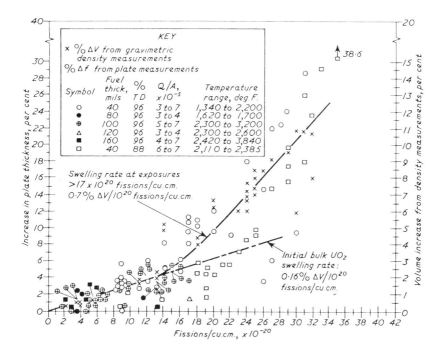

Fig. 6.3.   Volume increase versus burn-up - Bulk UO$_2$.
<u>Source</u>: R. C. Daniel *et al.*, WAPD-263 (1962).

Destructive examination includes the measurement of released fission gases;
sectioning and examination by optical metallography (including quantitative
metallography), scanning and transmission electron microscopy, electron
microprobe, and Auger electron spectroscopy; mechanical testing of the
cladding; and dissolution of fuel for burn-up analysis. The list is
incomplete because new analytical tools are being added as they become
available.

The results of all these tests constitute a form of forensic metallurgy,
whereby the life history and the cause of "death" of the fuel element is
deduced from the post-mortem, supplemented by the fabrication and reactor
operating history. This produces valuable data and explanations to feed back
to the other partners in the chain of development.

An example of this feedback process involves the phenomenon of pellet-cladding-
interaction, or PCI, which is the single most serious cause of cladding failure
in LWR elements. It assumed greater importance with the U.S. (Carter) policy
of no reprocessing and no plutonium recycling, because there is a premium on
fuel burn-up in a "once-through" cycle.

PCI, as revealed by post-irradiation examination, has two facets: *mechanical interaction* due to the stress-raisers at pellet-pellet interfaces and the ends of fuel cracks plus relative motion of fuel and cladding; and *chemical interaction* due to the stress-corrosion action of fission products on the cladding (iodine is believed to be the worst element in this respect). The two phenomena are interconnected, because the first place to fail by stress-corrosion cracking is the most highly stressed position. Furthermore, the volatile iodine most easily reaches the cladding along the fuel cracks and pellet-pellet interfaces. Careful post-irradiation examination of many LWR fuel elements has identified iodine in the vicinity of cladding cracks [9]. In other tests irradiated zircaloy tubing has been subjected to pressure tests in the presence of iodine vapour to confirm the role of this element [10]. A fairly comprehensive explanation of the phenomenon has been published and fuel element manufacturers have produced "fixes". One has been described in the open literature by the General Electric Company [11], who coated the inside wall of the zircaloy cladding with a thin layer of copper or titanium to provide a diffusion barrier against iodine and a lubricant to allow relative motion of fuel and cladding.

COMMERCIALIZATION

We have so far reviewed the progression in the development of a fuel element from a gleam in the reactor designer's eye to the generation of a fairly high level of confidence in the chosen design through the testing of a partial core loading. It cannot be emphasized too strongly that to arrive at this stage one may have to iterate many times; problems continually emerge that have to be solved satisfactorily. However, having overcome these problems the design must be evaluated in other terms.

First, there is the question of fabrication on a commercial scale. If the element represents a modification of an existing design, as in the case of coating the cladding inside wall to prevent pellet-cladding-interactions, it is highly probable that an existing commercial plant can accommodate the modification. If, on the other hand, the element is of a new design or is a radical modification of an existing design, a new plant or a new production line will be needed.

The cost of setting up a new fuel production line is very high and obviously must be justified in terms of the economic or financial return on the investment. Thus economic considerations will influence the choice of fabrication route. For example, in changing from mixed oxide to mixed carbide cores in LMFBRs one has, in principle, the choice between arc casting the carbide and powder pressing and sintering. However, the latter basically only involves the introduction of an extra step into a mixed oxide production line (a glovebox for carbothermic reduction). Hence, it makes better economic sense to follow this route. One should note that the economics of energy systems are not simple today, and a long-term shortage of uranium supply, exacerbated by a policy of restriction to once-through thermal reactors, could also influence decision-making on the choice of fabrication route. One might justify building an expensive metal fuel fabrication plant for a mixed spectrum reactor (FMSR) if it greatly improved our utilization of uranium.

In the course of designing and costing a fuel production plant one must be ever mindful of the need for high standards of quality assurance and quality control. Each element should be a complete replica of every other element and should reflect the wealth of knowledge acquired in the development process. Both operating economics and licensing criteria require this reproducibility of quality and performance.

Quality control is exercised through both destructive and nondestructive evaluations. Destructive methods generally involve wet chemical analysis of fuel stoichiometry and impurity levels, while nondestructive methods are used to measure fuel density, open and closed porosity and dimensions. Cladding is made from very well-characterized special alloy melts with destructive (metallographic) examination of the final tubular product (and welds) for grain size, inclusions and other flaws and nondestructive examination for flaws and the level of cold work.

When recycled fuel is fabricated (maybe one should say "if") there will be a need for remote, automated fabrication to avoid overexposure of the plant staff to gamma radiation. Already automated plants are being developed for fast reactor fuels because of the high gamma activity levels associated with the plutonium extracted from high burn-up fuels. The fabrication plant begins to resemble a hot cell, but is operated as automatically as possible.

In all of these considerations of commercial fuel element production one must include the question of interfacing with the remainder of the fuel cycle, i.e. mining, ore beneficiation, enrichment, reprocessing and waste handling. On the one hand one must be alert to possible deleterious effects from the front end of the cycle; an example from the past was the retention of excess fluorine from the enrichment process in UO_2 fuels, resulting in stress-corrosion cracking of the cladding in service. On the other hand, certain materials are difficult to dissolve or handle in reprocessing, e.g. carbide fuel and some refractory metal cladding materials. Certain materials make waste handling and plant maintenance more difficult, e.g. metals that acquire high levels of long-lived gamma activity, such as cobalt. Some efforts are currently being expended to minimize the solution and transport of long-lived gamma emitters from cladding materials to other parts of the primary circuit. Greater attention is being paid now to the integrity of irradiated fuel elements with respect to transportation to and behaviour in away-from-reactor storage sites. For example, the impact strength of irradiated cladding at or near normal climatic temperatures is of considerable interest.

Finally, one has to be fully aware of the methodology and criteria for licensing new cores [5,12]. The licensing procedure generally reviews all of the available data on fuel fabrication and performance (including off-normal or transient performance), relates this to the overall operation and performance of the plant and in particular uses modelling and probabilistic analysis techniques to answer "What if?" questions. An obvious series of questions are: "What are the consequences of various modes of cladding failure — on reactor operation and on district safety?" One such issue that has been well documented is the fuel densification process that was discovered through the observation of cladding collapse in irradiated fuel pins [13]. The cause of the collapse was the densification of the fuel, through disappearance of the fine sinter pores, causing shrinkage of the fuel column. The empty section of cladding collapsed under the coolant pressure. Concerns were expressed on the effects of the UO_2 column shrinkage on neutron leakage from a plane of the core, hence affecting core control, and on the effects of the cladding collapse on cladding integrity, thermal-hydraulics, etc. As described earlier, a quick "fix" was to prepressurize the cladding to prevent collapse [14], followed by a more permanent "fix" of eliminating fine pores from the UO_2. Both fixes were subjected to careful scrutiny by the NRC before licences were issued to operate these cores.

REFERENCES

1. Wrona, B. J., Roberts, J. T. A., Johanson, E., and Tuohig, W. D., *Nucl. Tech.* 20, 114 (1973).
2. Feraday, M. A., *et al.*, AECL-3111 (1969) and AECL-5028 (1975).
3. Domagala, R., Wiencek, T., Thresh, H. R. and Stahl, D., paper to International Meeting on the Development, Fabrication and Application of Reduced Enrichment Fuel for Research and Test Reactors, Argonne, IL (November 1980) to be published.
4. Walter, C. M., Olsen, N. J. and Hofman, G. L., *Proceedings of Symposium on Materials Performance in Operating Nuclear Systems*, CONF-730801 (NTIS), p. 181 (1973).
5. US Nuclear Regulatory Commission Standard Review Plan, Rev. 1, NUREG-75/087 or NUREG-0800 (1975).
6. Tokar, M., *et al.*, Nuclear Regulatory Report NUREG/CR 1818 (also PNL-3583) (1981).
7. Daniel, R. C., Bleiberg, M. L., Meieran, H. B. and Yeinscavich, W. B., U.S. Atomic Energy Commission Report WAPD-263 (1962).
8. Garzarolli, F., von Jan, R. and Stehle, H., *Atomic Energy Review*, 17 (1), 31 (1979).
9. Perrin, J. S., *et al.*, Electric Power Research Institute Report NP-812 (1978).
10. Yaggee, F. L., Mattas, R. F. and Neimark, L. A., Electric Power Research Institute Report NP-1155 (1979).
11. Rosenbaum, H. S., General Electric Report GEAP-23773-1 (1978).
12. NRC Regulatory Guide, 1.16 through Revision 5, Appendix A (1971 to 1979).
13. Meyer, R. O., *The Analysis of Fuel Densification*, NUREG-0085 (July 1976).
14. General Electric Licensing Topical Report NEDE-23786-1 (1978).

APPENDIX. SUPPLIERS OF FUEL, FUEL ELEMENT CLADDING, FUEL ELEMENTS AND FUEL SERVICES

Courtesy of the American Nuclear Society

Fuel Element Cladding

AGIP Nucleare S.p.A., Milan, Italy.
Atomergic Chemetals Corp., Plainview, N.Y.
The Babcock & Wilcox Co., Power Generation Group, Barberton, Ohio.
Cunnington & Cooper Ltd., Lancashire, U.K.
Decoto Nuclear Products, Yakima, Wash.
Fairey Engineering Ltd., Nuclear Div., Stockport, U.K.
Fansteel Inc., North Chicago, Ill.
Fine Tubes Ltd., Estover Works, Plymouth, Devon, U.K.
General Electric Co., Energy Systems & Technology Div., Advanced Reactor Systems Dept., Sunnyvale, Calif.
Hitachi Ltd., Tokyo, Japan.
Instel Corp., New York, N.Y.
Kawecki Berylco Industries, Inc., Reading, Pa.
Kobe Steel, Ltd., Tokyo, Japan.

Mannesmann Pipe & Steel Corp., Houston, Tex.
Mitsubishi Heavy Industries, Ltd., Tokyo, Japan.
Newburgh Engineering Company Ltd., Sheffield, U.K.
Noranda Metal Industries Ltd., Special Metals Div., Arnprior, Ontario, Canada.
Nuclear Components Inc., Great Barrington, Mass.
Pechiney Ugine Kuhlmann, Branche Nucleaire et Industries Diverses, Paris, France.
Rockwell International, Energy Systems Group, Canoga Park, Calif.
SICN-Societe Industrielle de Combustible Nucleaire, Neuilly sur Seine, France.
Sumitomo Metal Industries, Ltd., Tokyo, Japan.
Superior Tube Co., Norristown, Pa.
Teledyne Wah Chang Albany, Albany, Ore.

TI Steel Tube Div., Birmingham, U.K.

Tube Division of Kawecki Berylco Industries, Inc. Reading, Pa.

UOP Inc., Wolverine Div., Dearborn Hts., Mich.

Vallourec, Paris, France.

Westinghouse Canada Ltd., Atomic Power Div., Ontario, Canada.

Westinghouse Electric Corp., Pensacola Plant, Nuclear Component Div., Pensacola, Fla.

Westinghouse Electric Corp., Specialty Metals Div., Blairsville, Pa.

Fuel Elements, Fabricated

AGIP Nucleare S.p.A., Milan, Italy.

Asea-Atom, Vasteras, Sweden.

Atomergic Chemetals Corp., Plainview, N.Y.

The Babcock & Wilcox Co., Nuclear Materials and Mfg. Div., Apollo, Pa.

The Babcock & Wilcox Co., Power Generation Group, Barberton, Ohio.

Belgonucleaire S.A., Brussels, Belgium.

British Nuclear Fuels Ltd., Warrington, U.K.

Canadian General Electric Co., Ltd., Peterborough, Ontario, Canada.

CERCA (Compagnie pour l'Etude et la Realisation de Combustibles Atomiques), Creteil, Cedex, Paris, France.

Combustion Engineering, C-E Power Systems, Windsor, Conn.

Compagnie Generale des Matieres Nucleaires, Le Plessis-Robinson Cedex, France.

Compagnie Industrielle de Combustibles Atomiques Frittes (CICAF), Bollene, France.

Exxon Nuclear Co. Inc., Bellevue, Wash.

Framatome, Paris, La Defense, France.

General Atomic Co., San Diego, Calif.

General Atomic Co., TRIGA Reactor Div., San Diego, Calif.

General Electric Co., Energy Systems & Technology Div., Advanced Reactor Systems Dept., Sunnyvale, Calif.

General Electric Co., Nuclear Energy Group, San Jose, Calif.

Hitachi Ltd., Tokyo, Japan.

Institute for Resource Management, Inc., Bethesda, Md.

Japan Nuclear Fuel Co., Ltd., Kanagawa, Japan.

Kraftwerk Union AG, Muelheim a.d. Ruhr, FRG.

Lamco Industries, Inc., El Cajon, Calif.

Mitsubishi Heavy Industries, Ltd., Tokyo, Japan.

Newburgh Engineering Company Ltd., Sheffield, U.K.

Nuclear Fuel Industries, Ltd., Tokyo, Japan.

Nuclear Fuel Services, Inc., Rockville, Md.

Nukem GmbH, Hanau, FRG.

Pechiney Ugine Kuhlmann, Branche Nucleaire et Industries Diverses, Paris, France.

Rockwell International, Energy Systems Group, Canoga Park, Calif.

SICN–Societe Industrielle de Combustinle Nucleaire, Neuilly sur Seine, France.

"3-L" Filters Ltd., Cambridge (Galt), Canada.

Tokyo Shibaura Electric Co., Ltd., Tokyo, Japan.

Westinghouse Canada Ltd., Atomic Power Div., Ontario, Canada.

Westinghouse Electric Corp., Nuclear Commercial Operations Div.-NSSS & Nuclear Fuel, Pittsburgh, Penn.

Westinghouse Nuclear International, Brussels, Belgium.

Fuel Management Services

AGIP Nucleare S.p.A., Milan, Italy.

AMN Ansaldo Meccanico Nucleare S.p.A., Genoe, Italy.

The Babcock & Wilcox Co., Power Generation Group, Barberton, Ohio.

Bechtel Power Corp., San Francisco, Calif.

Belgatom, Bruxelles, Belgium.

Belgonucleaire S.A., Brussels, Belgium.

Black & Veatch Consulting Engineers, Kansas City. Mo.

British Nuclear Fuels Ltd., Warrington, U.K.

Burns and Roe, Inc., Oradell, N.J.

Canadian General Electric Co., Ltd., Peterborough, Ontario, Canada.

Ceco Consultants Ltd., Toronto, Ontario, Canada.

Combustion Engineering, C-E Power Systems, Windsor, Conn.

Compagnie Generale des Matieres Nucleaires, Le Plessis-Robinson Cedex, France.

Control Data Corp., Minneapolis, Minn.

Dames & Moore, Los Angeles, Calif.

Ebasco Services Inc., New York, N.Y.

Electrowatt Engineering Services Ltd., Thermal and Nuclear Energy Dept., Zurich, Switzerland.

Energy Inc., Idaho Falls, Id.

Framatome, Paris, La Defense, France.

General Electric Co., Nuclear Energy Group, San Jose, Calif.

Gibbs & Hill, Inc., New York, N.Y.

Gilbert/Commonwealth, Reading, Pa.

Hitachi Ltd., Tokyo, Japan.

E.R. Johnson Assoc., Inc., Vienna, Va.

Kraftwerk Union AG, Muelheim a.d. Ruhr, FRG.

Mitsubishi Atomic Power Industries, Inc., Tokyo, Japan.

Mitsubishi Heavy Industries, Ltd., Tokyo, Japan.

NIS Nuklear-Ingenieur-Service GmbH, Hanau, FRG.

Nuclear Associates International Corp, Rockville, Md.

Nuclear Assurance Corp., Atlanta, GA.

Nuclear Energy Services, Inc., NES Div., Danbury, Conn.

Nuclear Fuel Industries, Ltd., Tokyo, Japan.

Nuclear Power Services, Inc., Secaucus, N.J.

Nuclear Services Corp., Campbell, Calif.

Nukem GmbH, Hanau, FRG.

NUS Corp., Rockville, Md.

Pickard, Lowe and Garrick, Inc., Washington, D.C.

Robert H. Rahiser & Assoc. Inc., Columbia, S.C.

Sargent & Lundy, Chicago, Ill.

Scandpower A/S, Kjeller, Norway.

Scandpower, Inc., Bethesda, Md.

Sener S.A., Las Arenas Bilbao, Spain.

SNIA Techint-Tecnologie Energetiche
Avanzate S.p.A., Rome, Italy.
Southern Science Applications, Inc.,
Dunedin, Fla.
The S.M. Stoller Corp., New York, N.Y.
Stone & Webster Engineering Corp.,
Boston, Mass.
Studsvik Energiteknik AB, Nuclear
Technology Div., Nykoping, Sweden.
SwecoNuclear, Stockholm, Sweden S.
The Technical Research Centre of
Finland, Espoo, Finland.
Technology for Energy Corp.,
Knoxville, Tenn.
TECOP-3H, Suresnes, France.
TERA Corporation, Berkeley, Calif.
Torrey Pines Technology, San Diego,
Calif.
Transnuklear GmbH, Hanau, FRG.
University Computing Co., Dallas,
Tex.
Uranerzbergbau GmbH, Bonn, FRG.
Westinghouse Electric Corp.,
Nuclear Commercial Operations
Div.-NSSS & Nuclear Fuel,
Pittsburgh, Penn.

Fuel Materials

AGIP Nucleare S.p.A., Milan, Italy.
Allied Chemical Corp., Nuclear
Activities, Morristown, N.J.
Asea-Atom, Vasteras, Sweden.
Atomergic Chemetals Corp., Plainview,
N.Y.
The Babcock & Wilcox Co., Nuclear
Materials and Mfg. Div., Apollo,
Pa.
The Babcock & Wilcox Co., Power
Generation Group, Barberton,
Ohio.
British Nuclear Fuels Ltd.,
Warrington, U.K.
CERCA (Compagnie pour l'Etude et la
Realisation de Combustibles
Atomiques), Creteil, Cedex,
Paris, France.
Combustion Engineering, C-E Power
Systems, Windsor, Conn.
Compagnie Generale des Matieres
Nucleaires, Le Plessis-Robinson
Cedex, France.
Eldorado Nuclear Ltd., Ottawa,
Ontario, Canada.
Eldorado Nuclear Ltd., Port Hope,
Ontario, Canada.
Eurodif/Coredif, Bagneux, France.
Exxon Nuclear Co. Inc., Bellevue,
Wash.
Framatome, Paris, La Defense,
France.
General Atomic Co., San Diego,
Calif.
General Atomic Co., TRIGA Reactor
Div., San Diego, Calif.
General Electric Co., Energy Systems
& Technology Div., Advanced
Reactor Systems
Dept., Sunnyvale, Calif.
Mitsubishi Heavy Industries, Ltd.,
Tokyo, Japan.
Nuclear Assurance Corp., Atlanta, Ga.
Nuclear Fuel Services, Inc.,
Rockville, Md.
Nukem GmbH, Hanau, FRG.
Pechiney Ugine Kuhlmann, Branche
Nucleaire et Industries Diverses,
Paris, France.
Rockwell International, Energy
Systems Group, Canoga Park, Calif.
Separative Work Unit Corp. (SWUCO),
Gaithersburg, Md.
STEAG Kernenergie GmbH, Essen, FRG.
Teledyne Wah Chang Albany, Albany,
Ore.

TNS, Inc., Jonesboro, Tenn.
Uranerzbergbau GmbH, Bonn, FRG.
Urangesellschaft mbH & Co. KG,
Frankfurt/Main, FRG.
Urenco Ltd., Marlow, Buckinghamshire,
U.K.
Westinghouse Electric Corp., Nuclear
Commercial Operations Div.-NSSS
& Nuclear Fuel, Pittsburgh,
Penn.
Westinghouse Electric Corp,
Westinghouse Hanford Co.,
Richland, Wash.

Fuel Materials, Mixed-Oxide

AGIP Nucleare S.p.A., Milan, Italy.
Atomergic Chemetals Corp., Plainview,
N.Y.
The Babcock & Wilcox Co., Nuclear
Materials and Mfg. Div., Apollo,
Pa.
The Babcock & Wilcox Co., Power
Generation Group, Barberton,
Ohio.
Belgonucleaire S.A., Brussels,
Belgium.
British Nuclear Fuels Ltd.,
Warrington, U.K.
Combustion Engineering, C-E Power
Systems, Windsor, Conn.
Compagnie Generale des Matieres
Nucleaires, Le Plessis-Robinson
Cedex, France.
Framatome, Paris, La Defense, France.
General Electric Co., Energy Systems
& Technology Div., Advanced
Reactor Systems Dept.,
Sunnyvale, Calif.
Mitsubishi Heavy Industries, Ltd.,
Tokyo, Japan.
Nuclear Assurance Corp., Atlanta, Ga.
Nukem GmbH, Hanau, FRG.
Pechiney Ugine Kuhlmann, Branche
Nucleaire et Industries Diverses,
Paris, France.
Research Chemicals, Phoenix, Ariz.
Rockwell International, Energy Systems
Group, Canoga Park, Calif.

Fuel Materials, Plutonium

The Babcock & Wilcox Co., Power
Generation Group, Barberton, Ohio.
Belgonucleaire S.A., Brussels,
Belgium.
Compagnie Generale des Matieres
Nucleaires, Le Plessis-Robinson
Cedex, France.
Elektroschmelzwerk Kempten GmbH.,
Munchen, FRG.
General Electric Co., Energy Systems
& Technology Div., Advanced
Reactor Systems Dept., Sunnyvale,
Calif.
Nuclear Assurance Corp., Atlanta, Ga.
Rockwell International, Energy
Systems Group, Canoga Park, Calif.
SGN-Societe Generale pour les
Techniques Nouvelles, St.
Quentin-Yvelines Cedex, France.
STEAG Kernenergie GmbH, Essen, FRG.

Gloveboxes

ADC Medical Corp., Farmingdale, N.Y.
Belgonucleaire S.A., Brussels,
Belgium.
Cunnington & Cooper Ltd , Lancashire,
U.K.
Defense Apparel, Hartford, Conn.
Elwood Nuclear Safety, Buffalo, N.Y.

Excelco Developments, Inc., Silver
Creek, N.Y.
Fairey Engineering Ltd., Nuclear
Div., Stockport, U.K.
GEC Reactor Equipment Ltd., Leicester,
LE8 3LH, U.K.
General Dynamics, Electric Boat Div.,
Reactor Plant Services, Dept.
730, Groton, Conn.
Hamilton Industries, Two Rivers, Wis.
Kewaunee Scientific Equipment Corp.,
Special Products Div., Lockhart,
Tex.
Lamco Industries, Inc., El Cajon,
Calif.
Lintott Engineering Ltd., Horsham,
Sussex, U.K.
Luxatom Syndicat, Luxembourgeois
pour l'Industrie Nucleaire,
Steinfort, Luxembourg.
Mitsui Engineering & Shipbuilding
Co., Ltd., Tokyo, Japan.
Mohawk Industrial Supply, Inc.,
Manchester, Conn.
Murdock, Inc., Compton, Calif.
Murgue-Seigle, Meyzieu, France.
Newbrook Machine Corp., Silver Creek,
N.Y.
Niigata Engineering Co., Ltd., Tokyo,
Japan.
NTG Nukleartechnik GmbH und Partner,
Gelnhausen, FRG.
Nuclear Associates, Carle Place,
N.Y.
Nuclear Power Outfitters, McHenry,
Ill.
Nuclear Sources & Services, Inc.,
Houston, Tex.
Nukem GmbH, Hanau, FRG.
Overly Manufacturing Co., Greensburg,
Pa.
Pacific Steel Products Co., Inc.,
Seattle, Wash.
Pantatron Ltd., London, U.K.
Radiation Technology, Inc.,
Rockaway, N.J.
Robatel SLP1, Genas, France.
Safety and Supply Co., Rad-Safe Div.,
Seattle, Wash.
Sargent Industries, Central Research
Laboratories Div., Red Wing,
Minn.
S.I.P.P. Engineering, Nuclear Energy
Div., Bagnols-sur-Ceze, France.
Stainless Equipment Co., Englewood,
Colo.
Vacuum/Atmospheres Corp., Hawthorne,
Calif.

Glovebox Supplies

ADC Medical Corp., Farmingdale, N.Y.
Defense Apparel, Hartford, Conn.
Elwood Nuclear Safety, Buffalo, N.Y.
Euclid Garment Mfg. Co., Kent, Ohio.
General Scientific Equipment Co.,
Philadelphia, Pa.
Hamilton Industries, Two Rivers, Wis.
Kewaunee Scientific Equipment Corp.,
Special Products Div., Lockhart,
Tex.
Lintott Engineering Ltd., Horsham,
Sussex, U.K.
Luxatom Syndicat, Luxembourgeois
pour l'Industrie Nucleaire,
Steinfort, Luxembourg.
Metal Bellows Corp., Chatsworth,
Calif.
Mohawk Industrial Supply, Inc.,
Manchester, Conn.
NTG Nukleartechnik GmbH und Partner,
Gelnhausen, FRG.
Nuclear Power Outfitters, McHenry,
Ill.

Nuclear Sources & Services, Inc.,
Houston, Tex.
Nukem GmbH, Hanau, FRG.
Overly Manufacturing Co., Greensburg,
Pa.
Safety and Supply Co., Rad-Safe Div.,
Seattle, Wash.
Sargent Industries, Central Research
Laboratories Div., Red Wing,
Minn.
S.I.P.P. Engineering, Nuclear Energy
Div., Bagnols-sur-Ceze, France.
SOVIS, La Ferte/S/S Jouarre, France.
Vacuum/Atmospheres Corp., Hawthorne,
Calif.

Grids, Fuel Element

Allmetal Fabricators , Inc., New
York, N.Y.
CERCA (Compagnie pour l'Etude et la
Realisation de Combustibles
Atomiques), Creteil, Cedex,
Paris, France.
Fairey Engineering Ltd., Nuclear
Div., Stockport, U.K.
FIAT TTG S.P.A., Torino, Italy.
Kraftwerk Union AG, Muelheim a.d.
Ruhr, FRG.
Mitsubishi Atomic Power Industries,
Inc., Tokyo, Japan.
Neyrpic, Grenoble, France.
NPS Industries, Inc., Secaucus, N.J.
Nuclear Fuel Industries, Ltd.,
Tokyo, Japan.
Pechiney Ugine Kuhlmann, Branche
Nucleaire et Industries Diverses,
Paris, France.
Rockwell Engineering Company, Inc.,
Blue Island, Ill.
Teledyne Wah Chang Albany, Albany,
Ore.
TI Steel Tube Div., Birmingham, U.K.
Vereinigte Osterreichische Eisen und
Stahlwerke-Alpine Montan AG,
Nuclear Service Div., Linz,
Austria.

Graphite

Atomergic Chemetals Corp., Plainview,
N.Y.
British Nuclear Fuels Ltd.,
Warrington, U.K.
The Caborundum Co., Electro Minerals
Div., Niagara Falls, N.Y.
Compagnie Generale des Matieres
Nucleaires, Le Plessis-Robinson
Cedex, France.
Fairey Engineering Ltd., Nuclear
Div., Stockport, U.K.
Gravatom Industries Ltd., Gosport,
Hampshire, U.K.
Great Lakes Carbon Corp., Niagara
Falls, N.Y.
Huron Industries, Inc., Port Huron,
Mich.
Leico Industries, Inc., New York,
N.Y.
Luxatom Syndicat, Luxembourgeois
pour l'Industrie Nucleaire,
Steinfort, Luxembourg.
McMaster-Carr Supply Co., Chicago,
Ill.
NPS Industries, Inc., Secaucus, N.J.
Nuclear Shielding Supplies & Service,
Inc., Lauderdale-by-the-Sea, Fla.
Nuclear Shielding Supplies & Service
of Canada Ltd., Longueuil,
Quebec, Canada.
Pechiney Ugine Kuhlmann, Branche
Nucleaire et Industries
Diverses, Paris, France.

Stackpole Carbon Co., Carbon Div.,
St. Marys, Pa.
Union Carbide Corp., Carbon Products
Div., Sales Office, Chicago, Ill.

Manipulators, Remote

Alsthom-Atlantique, Engineering and
Mechanical Equipment Div., La
Courneuve Cedex, France.
Atcor - Washington, Inc., Farmington,
Conn.
Ateliers et Chantiers de Bretagne -
ACB, Nantes Cedex, France.
The Atlas Car & Manufacturing Co.,
Cleveland, Ohio.
BNP Nuclear Products, Indianapolis,
Ind.
Bristol Aerospace Ltd., Winnipeg,
Manitoba, Canada.
Cayuga Machine Inc., Depew, N.Y.
CEE-VEE Engineering Ltd., Bexhill-
on-Sea, Sussex, U.K.
Chem-Nuclear Systems, Inc., Columbia,
S.C.
Don L. Collins & Assoc., Glendale,
Calif.
Cunnington & Cooper Ltd., Lancashire,
U.K.
Dosimeter Corporation of America,
Cincinnati, Ohio.
Electrometer Corp., Cincinnati,
Ohio.
Fuji Electric Co., Ltd., Chiyodaku,
Tokyo, Japan.
GEC Reactor Equipment Ltd., Leicester,
LE8 3LH, U.K.
Gulton Industries, Inc., S.C.D. Div.,
Costa Mesa, Calif.
Hitachi Ltd., Tokyo, Japan.
Intercontrole, Rungis, France.
Luxatom Syndicat, Luxembourgeois pour
l'Industrie Nucleaire, Stainfort,
Luxembourg.
Maschinenfabrik Augsburg-Nurnberg AG,
Nurnberg, FRG,
Newburgh Engineering Company Ltd.,
Sheffield, U.K.
Nuclear Associates, Carle Place, N.Y.
Nuclear Systems Assoc., Brea, Calif.
Nucltec Products Inc., Fremont,
Calif.
Parameter, Inc., Consulting Engineers,
Elm Grove, Wis.
PaR Systems Corp., St. Paul, Minn.
Sargent Industries, Central Research
Laboratories Div., Red Wing, Minn.
Sundstrand Energy Systems, Rockford,
Ill.
TeleOperator Systems Corp., Bohemia,
N.Y.

Windows, Radiation-Shielded

ADC Medical Corp., Farmingdale, N.Y.
Atomic Products Corp., Center
Moriches, N.Y.
Chem-Nuclear Systems, Inc., Columbia,
S.C.
Don L. Collins & Assoc., Glendale,
Calif.
James Girdler & Co. Ltd., Bermondsey,
London, U.K.
Gravatom Industries Ltd., Gosport,
Hamshire, U.K.
R. V. Harty Div., Door-Man Mfg. Co.,
Royal Oak, Mich.
Industrial Engineering Works, Trenton,
N.J.
Jenaer Glaswerk Schott & Gen., Optics
Div., Optical Glass Sales Dept.,
Mainz, FRG.
Keene Corp., New York, N.Y.

Keene Corp., Ray Proof Div., Norwalk,
Conn.
Metex Thermal & Mechanical Group,
Edison, N.J.
Nippon Kogaku K.K., Tokyo, Japan.
Nuclear Associates, Carle Place,
N.Y.
Nuclear Lead Co., Inc., Oak Ridge,
Tenn.
Nuclear Pacific, Inc., Seattle,
Wash.
Nuclear Power Outfitters, McHenry,
Ill.
Nuclear Shielding Suppliers &
Service, Inc., Lauderdale-by-
the-Sea, Fla.
Nuclear Shielding Supplies &
Service of Canada Ltd.,
Longueuil, Quebec, Canada.
Nuclear Sources & Services, Inc.,
Houston, Tex.
Nuclear Supply Co., Washington, D.C.
Overly Manufacturing Co.,
Greensburg, Pa.
Panax Nucleonics Ltd., Redhill,
Surrey, U.K.
Ridge Instrument Co. Inc., Tucker,
Ga.
Schott Optical Glass inc., Duryea,
Pa.
SOVIS, La Ferte/S/S Jouarre, France.

Tubing - S Stainless
         SS Stainless, Seamless
         SG Steam Generator

Acieries du Manoir-Pompey, Pitres,
France. (S)
Alberox Corp., New Bedford, Mass.
Allegheny Ludlum Steel Corp.,
Pittsburgh, Pa. (S)
Al Tech Specialty Steel Corp.,
Watervliet, N.Y. (SS)
Aluminium Company of America,
Pittsburgh, Pa.
American Instrument Co., Silver
Spring, Md. (S)
Armco Inc., Advanced Materials Div.,
Baltimore, Md. (S)
Atomergic Chemetals Corp., Plainview,
N.Y.
Avica Equipment Ltd., Hemel
Hempstead, Herts., U.K. (S)
The Babcock & Wilcox Co., Power
Generation Group, Barberton,
Ohio. (S)
The Babcock & Wilcox Co., Tubular
Products Div., Beaver Falls,
Pa. (S, SS)
Babcock Power Ltd., London, U.K.
Bram Metallurgical-Chemical Co.,
Newtown, Pa.
Bristol Aerospace Ltd., Winnipeg,
Manitoba, Canada.
Capitol Pipe & Steel Products, Inc.,
Houston, Tex. (S, SS)
Carpenter Technology Corp., Tube Div.,
Union, N.J. (S)
Chase Nuclear (Canada) Ltd., Arnprior,
Ontario, Canada.
Chase Nuclear (U.S.) Div., Waterbury,
Conn.
Chicago Tube and Iron Co., Chicago,
Ill. (S, SS)
C.I.M.I. (Compagnia Italiana Montaggi
Industriali S.p.A.), Milano, Italy.
(S, SS)
Clark and Wheeler Engineering, Inc.,
Cerritos, Calif.
Combustion Engineering, C-E Power
Systems, Windsor, Conn. (S)
Commercial Fasteners Corp, New York,
N.Y. (S)
Compagnie Francaise d'Entreprises
Metaliques, Paris, France. (S)

Cunnington & Cooper Ltd., Lancashire,
U.K. (S)
Cyclops Corp., Sawhill Tubular Div.,
Sharon, Pa. (S)
Howard Dearborn, Inc., Fryeburg, Me.
(S, SS)
Defense Apparel, Hartford, Conn.
Elwood Nuclear Safety, Buffalo, N.Y.
Euclid Garment Mfg. Co., Kent, Ohio.
Fahramet Ltd., Orillia, Ontario,
Canada. (S)
Fansteel Inc., North Chicago, Ill.
Fine Tubes Ltd., Estover Works,
Plymouth, Devon, U.K. (S, SS)
Flextube Ltd., Hemel Hempstead,
Herts, U.K. (S)
Flextube S.A., Monte Carlo,
Principality of Monaco. (S)
Floy Tag & Mfg., Inc., Seattle,
Wash.
Gulfalloy, Inc., Houston, Text.
(S, SS)
Gulfalloy, Inc., Cranford, N.J.
(S, SS)
Gulfalloy, Inc., Norwalk, Calif.
(S, SS)
Guyon Alloys Inc., Wayne, Pa.
(S, SS)
High Voltage Engineering Corp.,
Burlington, Mass. (S)
Hitachi Ltd., Tokyo, Japan. (S)
Huntington Alloys, Inc., Huntington,
W. Va.
Intsel Corp., New York, N.Y. (S)
ITT Grinnell Corp., Providence, R.I.
ITT Harper, Morton Grove, Ill. (S)
Joliet Valves Inc., Minooka, Ill.
(S, SS)
Jones & Laughlin Steel Corp.,
Pittsburgh, Pa.
Kawecki Berylco Industries, inc.,
Reading, Pa. (S)
Kobe Steel, Ltd., Tokyo, Japan,
(S, SS, SG)
Kraftanlagen Aktiengesellschaft
Heidelberg, Heidelberg, FRG.
(S)
Kubota, Ltd., Tokyo, Japan. (S)
LaBarge Tubular Div., St. Louis, Mo.
(S)
Leland Tube Co., South Planfield,
N.J.
Liberty Equipment and Supply, Nuclear
Div., Kennewick, Wash. (S, SS)
Mannesmann Pipe & Steel Corp.,
Houston, Tex. (S)
Mannesmannrohren-Werke AG, Dusseldorf,
FRG.
McInnes Steel Co., Corry, Pa. (S)
McJunkin Corp., Charleston, W. Va.
(S, SS)
McMaster-Carr Supply Co., Chicago,
Ill. (S)
Metal Goods, St. Louis, Mo. (S, SS)
Mills Alloy Steel Co., Twinsburg,
Ohio. (S, SS, SG)
Samuel Moore and Co., Dekoron Div.,
Aurora, Ohio.
N.E.I. Clarke Chapman Power
Engineering Ltd., Gateshead,
Tyne and Wear, U.K. (SG)
Nikkiso Co., Ltd., Tokyo, Japan. (S)
Noranda Metal Industries Ltd.,
Special Metals Div., Arnprior,
Ontario, Canada. (S, SS)
NPS Industries, Inc., Secaucus,
N.J. (S, SS)
NRC Inc., Newton, Mass.
The Okonite Co., Ramsey, N.J.
Pacific Tube Co., Los Angeles,
Calif. (S, SS, SG)
Pechiney Ugine Kuhlmann, Branche
Nucleaire et Industries
Diverses, Paris, France.
Plymouth Tube Co., Inc., Winfield,
Ill. (S, SS)

Power & Engineered Products Co.,
Inc., South Plainfield, N.J.
Pressure Products Industries,
Warminster, Pa. (S)
Pressure Vessel Nuclear Steels,
Inc., Hillside, N.J.
Quartz Products Corp., Plainfield,
N.J.
Republic Steel Corp., Cleveland,
Ohio. (S)
Robert-James Sales, Inc., Buffalo,
N.Y. (S, SS)
Rollmet Inc., Irvine, Calif.
(S, SS)
Safety and Supply Co., Rad-Safe
Div., Seattle, Wash.
Sandvik Inc., Tubular Products Div.,
Scranton Works, Scranton, Pa.
(SS)
Sandvik Special Metals Corp.,
Kennewick. Wash.
R. & G. Schmoele Metallwerke GmbH
& Co. KG, Menden, FRG. (S, SS)
The Spencer Turbine Co., Windsor,
Conn. (S)
Stellite Div., Cabot Corp., Kokomo,
Ind.
Sumitomo Metal Industries, Ltd.,
Tokyo, Japan. (S)
Superior Tube Co., Norristown, Pa.
(S. SS)
Swepco Tube Corp., Clifton, N.J.
(S)
Teledyne Wah Chang Albany, Albany,
Ore.
Thypin Steel Co., Inc., Long
Island City, N.Y. (S)
Timet, Pittsburgh, Pa.
Tioga Pipe Supply Co., Inc.,
Philadelphia, Pa. (S)
TI Steel Tube Div., Birmingham,
U.K. (S, SS)
Tokyo Shibaura Electric Co., Ltd.,
Tokyo, Japan. (SG)
Trent Tube Div., Colt Industries,
East Troy, Wis. (S, SS)
Tube Division of Kawecki Berylco
Industries, Inc., Reading,
Pa. (SS)
Tubular Steel, Inc., Hazelwood,
Mo. (S)
Uniform Tubes, Inc., Collegeville,
Pa. (S, SS)
United Nuclear Corp., Washington,
D.C.
United States Steel Corp.,
Pittsburgh, Pa. (S)
UOP Inc., Wolverine Div., Decatur,
Ala.
UOP Inc., Wolverine Div., Dearborn
Hts., Mich.
Valin Corp., Sunnyvale, Calif.
Valley Steel Products Co., St. Louis,
Mo.
Vallourec, Paris, France. (S, SS)
Vereinigte Deutsche Metallwerke AG,
Tube Div., Duisburg, FRG.
(S, SS, SG)
Vereinigte Osterreichische Eisen und
Stahlwerke-Alpine Montan AG,
Nuclear Service Div., Linz,
Austria. (S)
Westinghouse Canada Ltd., Atomic
Power Div., Ontario, Canada.
Westinghouse Electric Corp.,
Specialty Metals Div.,
Blairsville, Pa. (SG)
Henry Wiggin & Co., Ltd., Hereford,
U.K. (SG)

Uranium Ore Concentrates

Chemical Separations Corp., Oak Ridge
Tenn.

Compagnie Generale des Matieres
Nucleaires, Le Plessis-Robinson
Cedex, France.
Denison Mines Ltd., Toronto, Ontario,
Canada.
Edlow International Co., Washington,
D.C.
Eldorado Nuclear Ltd., Ottawa,
Ontario, Canada.
Kerr McGee Nuclear Corp., Oklahoma
City, Okla.
Nuclear Assurance Corp., Atlanta,
Ga.
Nuclear Exchange Corp., Menlo Park,
Calif.
Pechiney Ugine Kuhlmann, Branche
Nucleaire et Industries Diverses,
Paris, France.
Rio Algom Ltd., Toronto, Ontario,
Canada.
Robatel SLP1, Genas, France.
STEAG Kernenergie GmbH, Essen, FRG.
The S.M. Stoller Corp., New York,
N.Y.
Transnuclear Inc., White Plains, N.Y.
Union Carbide Corp., Metals Div.,
New York, N.Y.
United Nuclear Corp., Washington,
D.C.
Uranerzbergbau GmbH, Bonn, FRG.
Urangesellschaft mbH & Co. KG,
Frankfurt/Main, FRG.

Uranium, Depleted

Atomergic Chemetals Corp., Plainview,
N.Y.
British Nuclear Fuels Ltd.,
Warrington, U.K.
Compagnie Generale des Matieres
Nucleaires, Le Plessis-Robinson
Cedex, France.
Reactor Experiments, Inc., San Carlos,
Calif.
TNS, Inc., Jonesboro, Tenn.
United Mineral & Chemical Corp.,
New York, N.Y.

Uranium Conversion

Allied Chemical Corp., Nuclear
Activities, Morristown, N.J.
Compagnie Generale des Matieres
Nucleaires, Le Plessis-Robinson
Cedex, France.
Comurhex, Paris, France.
Edlow International Co., Washington,
D.C.
Eldorado Nuclear Ltd., Ottawa,
Ontario, Canada.
Kerr McGee Nuclear Corp., Oklahoma
City, Okla.
Pechiney Ugine Kuhlmann, Branche
Nucleaire et Industries Diverses,
Paris, France.
The S.M. Stoller Corp., New York,
N.Y.
V/O Techsnabexport, Moscow, USSR.

Source: American Nuclear Society.

Uranium Enrichment Exchange Services

AGIP Nucleare S.p.A., Milan, Italy.
Compagnie Generale des Matieres
Nucleaires, Le Plessis-Robinson
Cedex, France.
Edlow International Co., Washington,
D.C.
Nuclear Assurance Corp., Atlanta, Ga.
Separative Work Unit Corp. (SWUCO),
Gaithersburg, Md.
STEAG Kernenergie GmbH, Essen, FRG.
Transnuclear Inc., White Plains, N.Y.

Uranium Enrichment Services

AGIP Nucleare S.p.A., Milan, Italy.
Compagnie Generale des Matieres
Nucleaires, Le Plessis-Robinson
Cedex, France.
Eurodif/Coredif, Bagneux, France.
Leybold-Heraeus GmbH, Hanau/Main,
FRG.
Separative Work Unit Corp. (SWUCO),
Gaithersburg, Md.
The S.M. Stoller Corp., New York,
N.Y.
Urenco Ltd., Marlow, Buckinghamshire
U.K.
V/O Techsnabexport, Moscow, USSR.

Uranium Exploration

Berge Exploration Inc., Denver,
Colo.
Compagnie Generale des Matieres
Nucleaires, Le Plessis-Robinson
Cedex, France.
EDA Instruments Inc., Toronto,
Ontario, Canada.
Golder Associates Inc., Kirkland,
Wash.
IRT Corp., San Diego, Calif.
Nuclear Assurance Corp., Atlanta,
Ga.
NUS Corp., Rockville, Md.
Pechiney Ugine Kuhlmann, Branche
Nucleaire et Industries
Diverses, Paris, France.
The S.M. Stoller Corp., New York,
N.Y.
Studsvik Energiteknik AB, Nuclear
Technology Div., Nykoping,
Sweden.
The Technical Research Centre of
Finland, Espoo, Finland.
Terradex Corp., Walnut Creek, Calif.
Urangesellschaft mbH & Co. KG,
Frankfurt/Main, FRG.
Western Systems, Inc., Scientific
Services Div., Evergreen, Colo.
Westinghouse Electric Corp., Nuclear
Instrumentation & Control Dept.,
Hunt Valley, Md.

CHAPTER 7

Experimental Techniques and Equipment

The process of testing fuel element designs and qualifying a specific design for commercial service is lengthy and expensive, comparable in many ways to the qualification of aerospace equipment.

The simplest part of the process is the fabrication of the fuel and the cladding and the measurement of their out-of-pile properties and behaviour. Even so this is not straightforward. First, one has to provide for nuclear materials accountability, criticality and safeguards. In the USA ^{239}Pu, ^{233}U and ^{235}U are accountable to within 0.1 g; no easy task when processing kilogram quantities through a long fabrication line. Second, one must protect the scientists and technicians from the toxicity and radioactivity of the fuel. This is done by the extensive use of gloveboxes, although small quantities may be handled in properly designed fume hoods. Obviously, proper radiation monitoring must be performed to check the exposure of personnel.

The general principle of glovebox design is that material is handled entirely inside airtight plastic and metal boxes. The atmosphere in the boxes may be air or an inert gas (usually helium), but in either case it is a cardinal rule that the pressure inside the box be lower than that outside so that any leakage is inwards. Elaborate gas circulation-pressure control systems are used that incorporate gas purification equipment and special exhaust filters to trap all particulate matter. Operations inside the boxes are performed by the use of long neoprene gloves that are hermetically sealed to circular openings in the glovebox wall (Fig. 7.1). The most risky part of the operation is the possible puncture of a glove. Usually one wears surgical gloves in addition to the stronger neoprene gloves to provide a second barrier. Gloves may be replaced without breaking the seals, the old gloves being dumped into the box.

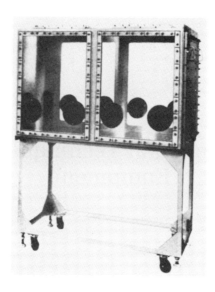

Fig. 7.1.   Standard glovebox mk la with stand. Source:
G. N. Walton (Ed.) "Glove Boxes and Shielded
Cells", Butterworths, 1958.

Transfers into and out of gloveboxes are performed through PVC transfer tubes
which are attached to the ends of the boxes. These transfer tubes are
thermally sealed at one end. To transfer material in and out of the box they
are heat-sealed in two or three places, a cut being made between these seals
to detach the object being transferred (see Fig. 7.2). Air locks have been
used, but are regarded as being less reliable than PVC transfer lines.

Obviously, the use of gloveboxes for nuclear fuel studies requires a lot of
training and experience. Operations take much more time than in the open
laboratory and must be planned much more carefully.

For fuel and fuel element research natural and low-enrichment UO_2 can be
handled without extensive use of gloveboxes, but they are required for all
other fuels including thorium. Typically, a line or "suite" of gloveboxes
is dedicated to fuel fabrication including (for ceramic fuels) powder
treatment, pressing, sintering, grinding and dimensional measurements. The
layout of a typical fuel fabrication facility is shown in Fig. 7.3. Other
characterization, such as density, pore size and pore distribution,
stoichiometry and grain size measurements, are performed in another line or
set of boxes. Property measurements and fuel element assembly, including
gas filling and welding, are carried out in other boxes. For metal fuels
melting and casting and other fabrication processes are performed in a suite
of boxes (Fig. 7.4). It is worth noting that the HFEP South facility at the
EBR-II reactor in Idaho was designed as a fuel recycle operation in which
irradiated metal fuel could be remelted (during which volatile fission
products left the fuel) and cast into rods for further irradiation. This
facility is inside heavy shielding for obvious reasons (Fig. 7.5). It points

up the fact that recycled fuel, even if it has passed through a wet
reprocessing plant, requires gamma shielding of the gloveboxes at a minimum
and more likely a totally remote operation akin to normal hot cell practice,
which will be discussed later.

To give some perspective, a multipurpose glovebox-equipped facility for fuel
and fuel element research would cost a *minimum* of $25 million to build in
1980. To make an experimental fuel pin or set of pins for an irradiation
experiment would take *at least* one year (and more likely two years) from
inception to delivery in 1980. Much of this time is involved with paperwork
— safety approvals, criticality statements, approval to irradiate in a
specific reactor, quality assurance, etc.

IN-PILE

As stated in the previous chapter, a vital part of fuel element design and
development is the testing of the element or rather a series of elements in
reactors. This testing goes through a progression, increasing in complexity
and realism up to the commercial reactor core loading.

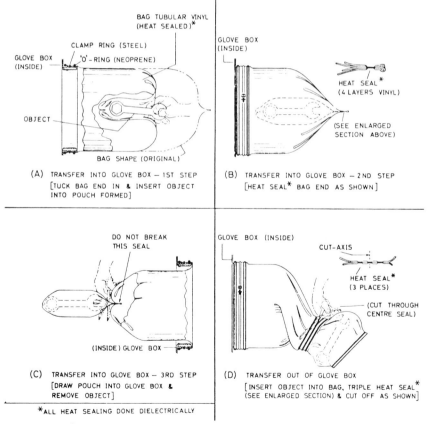

Fig. 7.2.   Research laboratory transfer procedure.
Source: G. N. Walton ibid.

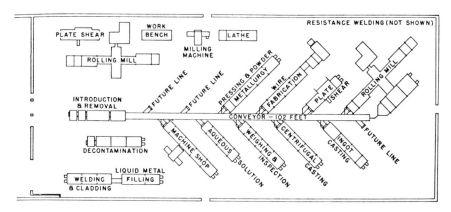

Fig. 7.3.   Fuel fabrication facility showing glovebox
arrangement.   Source: ANL.

The simplest testing, which is much less common today than it was in the
1950s and 1960s, is the study of fuel behaviour in research reactors.
Research reactors are generally small (in the 1 to 50 MWt range) built for a
number of purposes — neutron scattering, isotope irradiation and radiation
effects studies. Hence they are not optimized for radiation testing. The
neutron spectrum is generally soft, so that the fission rate varies from the
surface to the centre of a fuel sample due to self-absorption, i.e. the
burn-up of the sample varies considerably from surface to centre; the higher
the enrichment the worse the effect. The core height is generally no more
than 1 m and the length over which the flux is reasonably constant is much
less, hence one has to worry about axial as well as radial flux gradients.

Temperature control in research reactor tests is often difficult. The usual
philosophy is to design a test "capsule" or "vehicle" in which the fuel
sample or miniature fuel element is separated from the reactor coolant (or a
separate cooling circuit) by one or more carefully controlled gas-gaps
designed to produce a particular temperature change, the level of which may
be varied by the use of mixtures of gases of varying thermal conductivity
Figure 7.6 shows the cross-section of a capsule that was used in GETR to
irradiate fast reactor fuel pins. In addition, it is desirable to incorporate
an electric heater which will "fine-tune" the sample temperature through
thermocouples attached to the sample, and provide some heating to compensate
for reactor power changes.

There are no standard designs of capsules. Each experiment and each reactor
requires different characteristics. Examples of successful capsule experiments are measurements of unrestrained fuel swelling, fuel creep under
irradiation and mini-fuel element tests on advanced breeder reactor fuels [2].
An interesting variant is the swept capsule in which the fuel is exposed to
a flowing stream of helium that entrains fission gases and volatiles, like
iodine, to detectors so that fission product release rates can be measured.
An example of a fuel swelling rig for irradiation in a thermal reactor is
given in Fig. 7.7.

Fig. 7.4. Melting and casting unit. <u>Source</u>: ANL.

The next stage of complexity in fuel element testing takes place in test
reactors. It is not easy to differentiate between research and test reactors,
but one criterion is that the latter are usually designed for fuel element
testing combined with studies of coolant questions. Another criterion is
that test reactors can handle bundles of fuel elements, albeit often no more
than ten to twenty elements in a bundle.

Test reactors tend to fall into two categories — those that are really small
prototype reactors, such as EBR-II, DFR, Rapsodie, EBWR, VBWR, NRU and
perhaps Shippingport, and those that are large loop reactors, such as the FTR,
ETR and ATR in Idaho, BR-2 in Belgium and Melusine at Grenoble in France.
The popular DIDO-class reactor is basically a research reactor, but loops
have been installed in some of them.

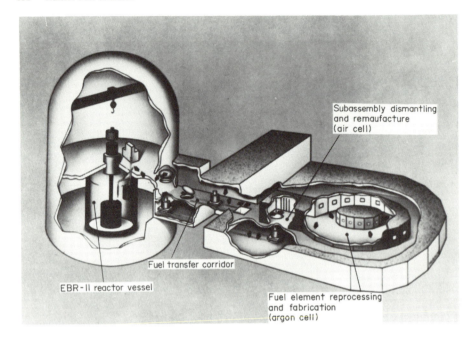

Fig. 7.5.   HFEF-South.   Source: ANL.

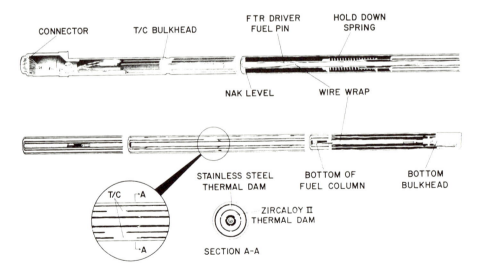

Fig. 7.6.   Schematic diagram of the fuel pin irradiation
capsule illustrating design features.
Courtesy: Hanford Engineering Development
Laboratory.

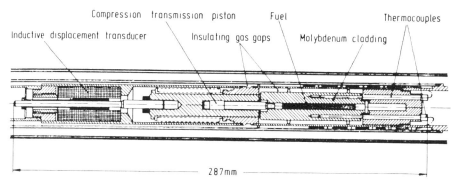

Fig. 7.7.   Swelling capsules.   Source: H. Zimmermann,
KfK, Karlsruhe, Germany.

A loop is an independently-cooled circuit in which one can test coolants
other than that used in the host reactor and perform "risky" experiments such
as run-beyond-cladding-breach when the debris is confined to the loop. Most
loops remain in the reactor for many test cycles, being equipped for the
loading and unloading of test bundles.

Transient test reactors form a subset of the test reactor class. Irradiated
fuel elements, usually in clusters, are tested in loops in the core of the
transient reactor, the whole loop being removed for post-test examination
because they usually contain a lot of debris.

Large loops, such as those in the ATR and BR-2 reactors, are virtually small
reactor coolant circuits with their own pumps, heat exchangers and coolant
chemistry monitoring and control systems. They are very expensive to build,
install and operate, but are obviously much cheaper than building a reactor
or risking certain operations in a whole reactor core.

A relatively simple example of a loop for insertion in a test reactor is the
CLIRA (Closed Loop In-reactor Assembly) for FFTF (Fig. 7.8) [3]. This is a
flowing sodium loop for insertion in a core or blanket subassembly position.
It has its own pumps, intermediate heat exchanger and surge tank, the coolant
flow being independent of the reactor coolant flow.

Subassemblies for insertion in loops or in the cores of test reactors are
often experimental in nature. For example, a number of instrumented
subassemblies have been built and inserted in the EBR-II core. These were
designed to measure features that are not possible in the normal driver
subassembly, such as fuel and cladding temperatures or cladding creep rates
at controlled temperatures. An example is the XX07 Test Assembly which
measured the coolant flow, temperature and flux levels during the irradiation
of 57 uranium-5 wt % fissium fuel elements to 2.9 at % burn-up [4]. The
subassembly incorporated two permanent magnet flowmeters, 23 thermocouples
(10 on fuel elements and 13 in the coolant) and two self-powered neutron
detectors. A sketch of this subassembly is shown in Fig. 7.9. Its design,
construction, assembly and operation was a complex process extending over
several years. It provided unique and valuable data on the conditions in the
core during reactor start-ups and shutdowns, steady state operation at various
power levels, reactor scrams, coolant flow coast downs and natural convective
flow conditions.

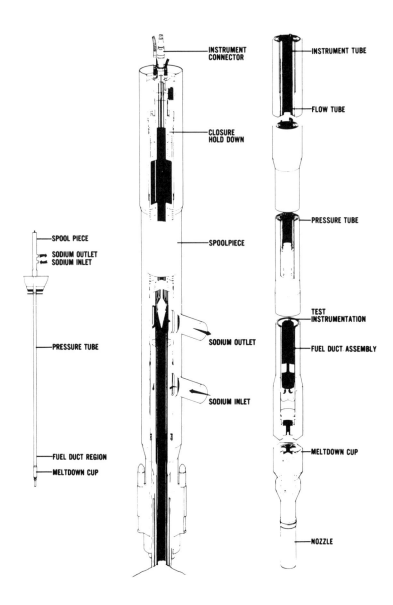

Fig. 7.8.   CLIRA loop for FFTF.   Courtesy: Hanford
Engineering Development Laboratory.

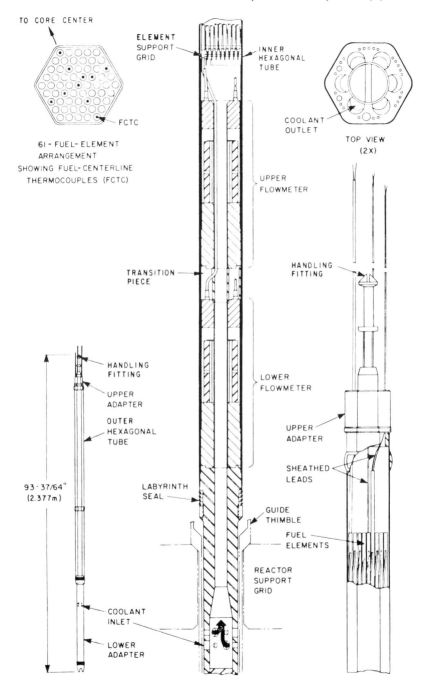

TO CORE CENTER

FCTC

61-FUEL-ELEMENT
ARRANGEMENT
SHOWING FUEL-CENTERLINE
THERMOCOUPLES (FCTC)

ELEMENT
SUPPORT
GRID

INNER
HEXAGONAL
TUBE

COOLANT
OUTLET

TOP VIEW
(2X)

UPPER
FLOWMETER

TRANSITION
PIECE

HANDLING
FITTING

HANDLING
FITTING

UPPER
ADAPTER

OUTER
HEXAGONAL
TUBE

LOWER
FLOWMETER

UPPER
ADAPTER

93-37/64"
(2.377m)

LABYRINTH
SEAL

SHEATHED
LEADS

GUIDE
THIMBLE

COOLANT
INLET

FUEL
ELEMENTS

LOWER
ADAPTER

REACTOR
SUPPORT
GRID

Fig. 7.9.   Instrumented subassembly XX07.   <u>Source</u>: ANL.

This subassembly was withdrawn from the reactor because it was suspected
that one or more fuel elements had developed leaks in the cladding. This
was verified by subsequent post-irradiation examination.

The test procedures described to date enable the fuel element and core
designers to optimize a fuel element design, through a series of refinements
and iterations of design and test. The final stage of development comes when
a fuel element design is chosen for a power reactor. Usually the power
reactor core is larger than that of the test reactors, so that the fuel
elements are longer and the environment in the core is different in terms of
flux and temperature gradients, control rod placement, neutron flux levels
and spectra as a function of time and operating mode, etc. Hence, it is
necessary to test single subassemblies and partial core loadings in a
representative power reactor core before going to a full core loading. In
the development of most thermal power reactors the earliest power plants
were small (in the range 50-200 MWe) compared to the 1000 MWe size that is
now standard. Hence there was some extrapolation of technology in that
scale-up process as discussed in Chapter 8.

Power reactor testing generally does not involve unusual instrumentation or
operation; the new assemblies are treated like the old ones except that they
are marked down for special attention at the post-testing phase. The bulk
of post-test examinations of power reactor fuel elements consist of nondestructive examinations under water in the reactor storage pool. A limited
number of assemblies or elements are subjected to destructive examination.
Generally, these are the ones that failed or those that are experimental in
nature.

Without going into detail, it is worth noting that statistical testing of
fuel elements is very important. The statistical significance of a single
subassembly test is not very high when it comes to designing a new reactor
core. The characteristic of fuel element failure statistics is that they
follow a U-shaped curve of numbers of elements failed as a function of
burn-up (Fig. 1.6). The curve is positive at low burn-ups, representing
"infant mortality", usually due to inadequate quality assurance resulting in
weld failures. The position of the sharp upturn in the curve is important
to establish, because this has strong implications on the fuel cycle economics.
Alternatively, it sets the standard for improvement by the fuel element
designer and developer.

POST-IRRADIATION EXAMINATION

Since the examination of fuel elements in storage pools has already been
mentioned, this section will be confined to hot-cell examinations.

A hot cell is a shielded box or room in which irradiated materials are
remotely manipulated. This can range in complexity from a small box made up
of lead bricks, with long tongs mounted in lead balls, to a vast concrete
hall in which many complex operations are performed by means of electrically
operated master-slave manipulators or radio-controlled robots.

The operations performed in hot cells include:

1. The turn-round of reactor experiments. For example, a subassembly is
    removed from the core of a test reactor into a hot cell, where several
    elements are removed for examination and are replaced by fresh elements.
    Some *interim* nondestructive examination may be performed on the elements
    being returned to the reactor.

2. Nondestructive examination at end of life. This includes X-ray, gamma-ray or neutron radiography, gamma-scanning, profilometry, etc.
3. Destructive examination down to fine detail.

The main objective of activites 2 and 3 is to determine how the fuel element behaved when it was in the reactor; sometimes described as "forensic metallurgy" because of the close analogy to the role of a coroner in determining the cause of death by means of an autopsy.

The characteristics of hot cells can be illustrated by reference to those with which the author is most familiar, i.e. at Argonne National Laboratory and at AERE Harwell. The overall philosophy at both laboratories is similar, i.e. the use of a large, engineering-oriented facility to handle large pieces of hardware (especially loops) and a smaller, metallurgy-oriented facility to examine the parts of the loops or subassemblies in considerable detail.

The Hot Fuel Examination Facility-North or HFEF/N at Argonne-West in Idaho is the largest and (in 1980) the most modern inert-atmosphere, alpha-gamma hot cell in the United States. Its purpose is to handle subassemblies from EBR-II (and possibly FTR), and sodium loops from TREAT and SLSF. The complex shown in Fig. 7.10 is built of high density concrete with lead-glass windows with Argonne-designed master-slave manipulators at 21 work stations, plus three bridge-mounted electromechanical (heavy duty) manipulators and three 5-ton cranes. The cell structure consists of a 56 m^2 floor area air-filled decontamination cell connected to a 195 m^2 area cell filled with argon controlled to less than 50 ppm oxygen and water vapour. Atmosphere control is important because sodium-filled loops are opened up in this cell. Attached to the cell is a TRIGA-FLIP reactor rated at 250 kW for neutron radiography of articles up to 3 m long × 16.5 cm diameter. The largest items handled up to the present in the HFEF/N are the $\sim$9 m long SLSF loops (Fig. 7.11).

Typical operations in HFEF/N are subassembly turn-round, loop disassembly, and the examination of EBR-II subassemblies and fuel elements. The techniques available for the latter task are a precision gamma-scanner, an eddy current tester for cladding integrity, profilometry, optical and scanning electron microscopy and neutron radiography. The facility is operated by a group of approximately 100 engineers and technicians, excluding supporting services.

Parts of loops and whole experimental fuel elements may be shipped from HFEF/N to a hot cell at Los Alamos or to a basically similar hot cell at Argonne-East in Illinois. The latter is known as the Alpha-Gamma Hot Cell Facility or AGHCF (Fig. 7.12). This is also built of high density concrete with zinc-bromide/glass windows and remote manipulators. It is much smaller in size than HFEF/N, having been designed for detailed examinations. The cell has two levels of atmosphere control, the higher purity are containing nitrogen with less than 50 ppm oxygen and water vapour, the lower purity area containing nitrogen with $\sim$0.25 wt % oxygen.

The high purity atmosphere side of the cell is a destructive examination area in which fuel elements or loops are sectioned for metallographic examination and for examination by the electron microprobe, scanning and transmission electron microscopes and an Auger spectrometer that are located in rooms adjacent to the cell (Fig. 7.13a). Each instrument is individually shielded for alpha and gamma activity. A computerized Quantimet analyzer is used for counting features in optical and electron micrographs such as pore and bubble sizes and numbers.

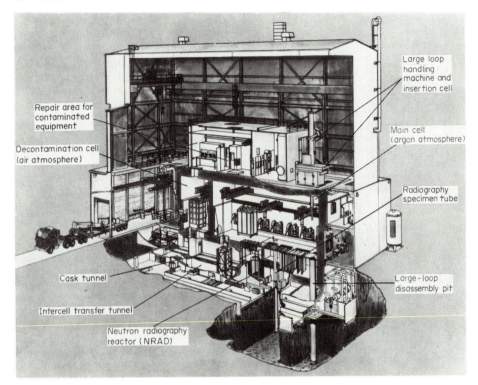

Fig. 7.10.   HFEF-North.   <u>Courtesy</u>: ANL.

The lower purity side of the cell contains the receiving, maintenance and nondestructive examination areas that include gamma scanning, profilometry (Fig. 7.13b), eddy current inspection, fission gas sampling, together with in-cell experiments such as direct electrical simulation of fission and mechanical tests on irradiated cladding.

Adjacent to the main cell are several shielded gloveboxes with simple tongs to operate such processes as ultrasonic coring of ceramic fuel pellets and chemical thinning of samples for electron microscopy.

The Argonne complex of hot cells also includes a small beta-gamma cell for mechanical testing of cladding and structural materials and a Chemistry hot cell which essentially provides 12 separate hot laboratories connected by radio-controlled transfer carts and robots. Cells are equipped for the chemical analysis of radioactive materials, additional metallography, electron and ion probe studies and irradiation of samples with a 80 kilocurie ^{60}Co gamma source.

Fig. 7.11.   SLSF loop being loaded into HFEF-North hot
cell.   Courtesy: ANL.

Fig. 7.12. Alpha-gamma hot cell facility. <u>Courtesy</u>: ANL.

The arrangements at Harwell are rather similar to those at Argonne, with a
High Activity Handling Building, which provides a general service for
dismantling irradiation rigs and loops, and a Metallurgical Post-irradiation
Facility. While the latter has a concrete cell with master-slave manipulators,
most of the work is carried out in free-standing shielded gloveboxes, one of
which is shown in Fig. 7.14. These are made of thick lead blocks stacked
around stainless steel gloveboxes with air bearing lead ball joints through
which the handling tongs are passed (Fig. 7.15).

Obviously there are trade-offs between the large cell and shielded glovebox
concepts. Transfers are easier in the former, repairs and decontamination
are easier in the latter. It is also easier to add new techniques by
installing them and testing them nonactively in shielded gloveboxes than to
do this in a large, operating cell.

Economics are also a powerful factor. To build a cell comparable to HFEF/N
starting today (1980) would cost at least $40 million, while a shielded
glovebox could be built for less than $0.5 million including the cost of the
lead and hook-up to utilities. However, if the objective is to examine fuel
elements from 1000 MWe commercial thermal reactors, one has to go to the $40
million scale cell simply to handle the large-scale operations that are
involved. On the other hand, small-scale research, such as fuel irradiations
in research or test reactors, can be accommodated in several shielded gloveboxes that house gas sampling, density and metallographic equipment.

Fig. 7.13(a)   The AGHCF's alpha-gamma containing scanning
              electron microscope is shown recording the
              x-ray spectra of an irradiated specimen
              contained in the alpha glovebox shown at
              the rear.

Fig. 7.13(b) Profilometer for measuring external
              dimensions of fuel pins.

In this chapter it is only intended to give an outline of what is involved
in radiation performance testing. The location of major facilities are
listed in Table 7.1 and a partial listing of U.S. suppliers of the specialized
equipment and services is given in the Appendix to Chapter 6.

TABLE 7.1. Centers for Fuel Element Research

Nation	Centre	Reactors	Hot cells	Specialties	Principal fuel element specialists
USA	Argonne National Laboratory Argonne, Illinois	None	AGHCF	LMFBR fuel performance, modelling and properties. LWR fuel failure mechanisms.	B. Frost C. Johnson J. Lambert L. Neimark F. Nichols R. Weeks
"	Argonne National Laboratory Idaho Falls, Idaho	EBR-II TREAT	HFEF	Metal fuel fabrication and performance. Post-test examination.	P. Bacca B. Seidel L. Walters
"	Oak Ridge National Laboratory Oak Ridge, Tennessee	ORR HFIR	TRU Others	HTGR fuel development. Sol-gel fabrication. Research reactor fuels. Reprocessing.	J. Cunningham M. Feldman L. Lotts T. Washburn
"	Los Alamos Scientific Laboratory Los Alamos, New Mexico		MET Cells	Carbide fabrication, properties and performance. General hot cell examination.	J. Barner J. L. Green T. Latimer
"	Hanford Engineering Development Laboratory, Richland, Washington	FFTF	FMEF	Fast reactor fuel element development and testing.	C. Cox E. Evans J. Laidler R. Leggett W. Roake
"	Battelle Pacific Northwest Laboratory, Richland, Washington	—	—	LWR fuel element performance and safety. Nuclear waste.	M. Freshley S. Goldsmith
"	Battelle Columbus Laboratory Columbus, Ohio	—	Large general purpose	LWR fuel examination. Fabrication of fuels. Waste management.	A. Bauer D. Keller J. Perrin

Nation	Centre	Reactors	Hot cells	Specialties	Principal fuel element specialists
USA	EG&G, Idaho National Engineering Laboratory, Idaho Falls, Idaho	ETR ATR PBF LOFT	TAN	LWR fuel performance – especially safety.	J. Crocker T. Dearien P. MacDonald
"	Westinghouse Advanced Reactors Division, Waltz Mill, Pennsylvania	–	–	LMFBR fuel performance.	A. Biancheria A. Boltax B. Harborne P. Levine P. Murray
"	Westinghouse Bettis Atomic Power Laboratory, Pittsburgh, Pennsylvania	–	–	Naval reactor fuel elements. Light-water breeder reactor.	J. Fong S. Harkness
"	Westinghouse Commercial Reactor Division, Pittsburgh, Pennsylvania	–	–	LWR fuel elements.	L. Boman K. Jordan E. Roberts
"	General Electric Co. Vallecitos Laboratory	GETR	General purpose	LMFBR fuel development.	M. Adamson E. Aitken
"	General Electric Co. Commercial Reactors, San Jose & Sunnyvale, California	–	–	LWR fuel development.	S. Armijo W. Baily H. Klepfer R. Proebstle
"	General Electric Knolls Atomic Power Laboratory, Schenectady, New York	–	–	Naval Reactor fuel elements.	E. F. Koenig
"	Combustion Engineering, Windsor Connecticut	–	–	LWR fuel elements.	M. Andrews W. Chernock R. Duncan S. Pati

Nation	Centre	Reactors	Hot cells	Specialties	Principal fuel element specialists
USA	Babcock & Wilcox, Lynchburg, Virginia	–	General purpose	LWR fuel elements.	T. Papazoglou M. Montgomery J. Tulenko
"	Exxon Nuclear, Richland, Washington	–	–	LWR fuel elements.	O. Kruger K. Merckx
"	Rockwell International, Canoga Park, California	–	–	LMFBR fuel elements. Research reactor fuel elements.	H. Pearlman
"	Stoller Corporation	–	–	Fuel performance and fuel cycle analysis.	K. Lindquist A. Strasser
"	Department of Energy, Germantown Maryland	–	–	Planning and analysis of reactor programmes.	W. Bennett K. Magnus R. Neuhold
"	Nuclear Regulatory Commission, Silver Spring, Maryland	–	–	Reactor safety analysis and licensing.	W. Johnston R. Meyer M. Picklesheimer L. Rubinstein
"	Electric Power Research Institute, Palo Alto, California	–	–	LWR fuel element performance analysis.	F. Gellhaus H. Ocken J. T. A. Roberts
UK	AERE, Harwell, Didcot, Oxon	DIDO PLUTO	General purpose	LMFBR, gas-cooled and LWR fuel element research.	R. Bellamy D. Harries J. Sayers

Nation	Centre	Reactors	Hot cells	Specialties	Principal fuel element specialists
UK	Risley Nuclear Laboratory, Warrington	-	-	Fuel element performance, design and analysis (all types).	K. Bagley, J. F. W. Bishop, E. Hicks, D. Linning, J. Pounder
"	Springfields Nuclear Laboratory, Preston, Lancs	-	-	Fuel element fabrication, testing and analysis.	J. Gittus, D. Locke, D. Pickman
"	Windscale Laboratory, Seascale	WAGR Calder Hall	General purpose	Fuel element development and testing.	V. W. Eldred, A. Garlick, J. Horspool, J. Nairn, A. Parkinson
"	Dounreay Experimental Reactor Establishment, Thurso, Scotland	PFR	General purpose	LMFBR fuel development and testing.	W. Batey, B. Edmonds, K. Swanson
"	Winfrith Atomic Energy Research Establishment, Winfrith, Dorset	SGHWR	General purpose	Fuel element testing and examination.	R. Cornell, S. A. Cottrell
"	CEGB Laboratory, Berkeley, Gloucestershire	Berkeley Magnox Reactor	General purpose	Fuel element development and examination	B. Edmondson, J. Harris, R. Hesketh, J. Waddington
"	British Nuclear Fuels Limited, Risley, Springfields & Windscale	WAGR Magnox Stations	General purpose	Fuel production and testing.	T. Heal, L. Raven
West Germany	KFA, Jülich	AVR	General purpose	HTGR fuel element research.	

Nation	Centre	Reactors	Hot cells	Specialties	Principal fuel element specialists
West Germany	KFK, Karlsruhe	FR-2 KNK	General purpose	LMFBR fuel element development.	G. Karsten, K. Kummerer, D. Vollath
"	Kraftwerk Union, AG Erlangen	-	-	LWR fuel element design, development and testing.	F. Garzarolli, R. Holzer, W. Kaden, H. Stehle
"	Nukem GMBH, Hanau	-		Fuel element fabrication.	
Austria	Reactor Center Seibersdorf	ASTRA		HTGR fuel element research.	
Canada	AECL, Chalk River Nuclear Laboratory, Deep River, Ontario	NRU NRX	General purpose	CANDU reactor fuel element research and testing.	B. Cox, A. Bain, D. Hardy, R. MacEwan, M. J. F. Notley, J. A. L. Robertson, J. Wood
"	AECL, Whiteshell Nuclear Laboratory, nr Winnepeg, Manitoba	WRL	General purpose	CANADU and other reactor fuel development.	
Denmark	RISØ Research Establishment	DR-3	General purpose	LWR fuel element development.	H. Carlsen, N. Hansen, P. Knudsen
EURATOM	Transuranium Institut, Karlsruhe, W. Germany	-	General purpose	Research on reactor fuel materials.	H. Blank, R. Lindner, Hj Matzke, C. Ronchi

Nation	Centre	Reactors	Hot cells	Specialties	Principal fuel element specialists
EURATOM	ISPRA Research Institute, Italy	ISPRA-1 ECO ESSOR		General research on fuels and cladding.	M. Schüle
"	Petten Research Establishment, Netherlands			Cladding development carbide fuel studies.	E. VandenBemden
Belgium	CEN Mol	BR-2 BR-3	General purpose	Fuel and cladding loop tests.	
"	Belgonucleaire, Brussels, Belgium				H. Bairiot P. Bouffiot R. Godesar M. Guyette
France	CEA, Saclay Laboratory, nr Paris	EL-2 EL-3 Osiris	General purpose	Research on fuels and cladding.	P. Blanchard P. Delpeyroux R. Lallement P. Mustelier
"	CEA, Fontenay-aux-Roses nr Paris	Triton Minerve	–	Plutonium studies.	P. Combette R. Pascard J. Roualt F. Sebilleau
"	CEA, Cadarache	Rapsodie Cabri Pegase	General purpose	LMFBR fuel fabrication and testing (normal and transient).	J. Melis
"	CEN, Grenoble	Melusine Siloé	General purpose	Materials research.	J. Blin

Nation	Centre	Reactors	Hot cells	Specialties	Principal fuel element specialists
France	EDF, Chinon	3 gas cooled reactors	General purpose	Gas-cooled reactor development.	
"	Framatome, Courbevoie	–	–	Reactor manufacturer.	
"	SICN, Neiully	–	–	Fuel element fabrication.	
Italy	CNEN, Cassaccia			Research on fuels and materials.	A. Pulacci M. Gabaglio
"	AGIP Nucleare, Milan		–	Fabrication of fuels.	
"	Combustibili-Coren Saluggia (Vercelli)		–	Fabrication of fuels and fuel elements.	
"	ENEL, Rome		–	State utility. All aspects of reactor development.	
Japan	Japan Atomic Energy Research Institute, Oharai & Tokai	JRR-2,3,4 JMTR JPDR 1&2	General purpose	Research on nuclear fuels.	S. Nomura J. Shimokawa H. Watanabe
"	Power Reactor & Nuclear Fuel Development Corp. (PNC), Oarai	Joyo Fugen ATR	General purpose	Fast reactor development.	M. Katsuragawa K. Noro K. Uematsu
"	Hitachi Ltd., Tokyo	HTR	–	Reactor manufacturer.	
"	Japan Nuclear Fuel Co., Kanagawa			Nuclear fuel R&D and fabrication.	
"	Mitsubishi Atomic Power Industries, Tokyo			Reactor manufacturer.	

Nation	Centre	Reactors	Hot cells	Specialties	Principal fuel element specialists
Japan	Nuclear Fuel Industries Ltd., Tokyo			Nuclear fuel R&D and production.	
Netherlands	Reactor Centrum Nederland (RCN) Petten NH	HFR		Irradiation experiments.	R. Blackstone
Norway	Institut for Atomenergi Kjeller	JEEP Halden (OECD)	General purpose	LWR fuel element irradiation.	S. Aas K. Videm
	SCANDPOWER Kjeller & Halden	–	–	Fuel technology.	E. Rolstad
So. Africa	South African Atomic Energy Pelindaba	SAFARI			J. P. B. Hugo N. Pienaar C. vander Walt
Sweden	AB ATOMENERGI Stockholm & Studsvik Center	R2	General purpose	Development, fabrication and testing of water reactor fuel elements.	S. Djurle M. Grounes H. Mogard U. Runfors
	ASEA Nuclear Power Dept., Vasteras	–	–	Development, fabrication and testing of fuel assemblies.	K. Hannerz S. Junkrans O. Värnild
Switzerland	EIR (Institute for Reactor Research) Wurenlingen	DIORIT	General purpose	Mainly carbide fuel development and testing.	K. Bischoff L. Smith R. Stratton
	Brown-Boverei & Cie Baden	–	–	Reactor manufacturer.	

Nation	Centre	Reactors	Hot cells	Specialties	Principal fuel element specialists
USSR	Institute of Atomic Reactors, Melekess	SM2 BOR-60	General purpose	Fuels and materials testing. Fast reactor fuels.	O. D. Kazachkovsky
"	Institute of Atomic Energy – I.V. Kurchatov, Moscow	MIR MR IRT WWR-S			
"	Institute of Atomic Energy A.F. Joffe, Leningrad	WWR-M		Reactor materials research.	
"	Physical Institute of Georgian Academy of Sciences, Tblisi	IRT-2000		Reactor materials research.	
"	Kiev Physics Institute, Kiev	WWR-M		Materials research.	N. Platov
"	Soviet Atomic Laboratory Obninsk	BFS BRI→5 WWR-T$_s$			I. Golovnin
"	Physical Engineering Institute Kharkov			Nuclear reactor components.	

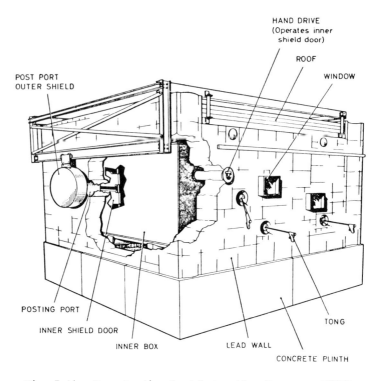

Fig. 7.14.   Freestanding lead hot cell.   <u>Courtesy</u>: UKAEA,
Harwell, England.

It remains to discuss the strategy in the use of these complex and expensive
facilities. The key to this is the cadre of fuel element design and
development experts who, using the principles described in previous chapters,
define a fuel element and subassembly concept. Having done that they must
lay out a suitable test programme that includes single pin and bundle tests
under steady state, off-normal and accident conditions. At that stage the
ball passes to the fuel element fabricators, the test engineers, reactor
operations personnel and the hot cell operators, each of whom has a vital
role to play in making fuel elements, irradiating them under known conditions
and examining them. The design group receives a mass of data: unirradiated
fuel element vital statistics, reactor operations data for the duration of
the test, and post-irradiation examination data. From this they must understand and evaluate the fuel performance against criteria which were set at
the beginning and against fuel element code predictions. This may well
provoke the need for further post-irradiation tests to throw light on specific
problems. This will probably lead to a refinement in the design to obviate
the problems encountered in the test and will require another test to check
out the refinement. Since this is a very expensive and time-consuming process,
the fewer the iterations the better.

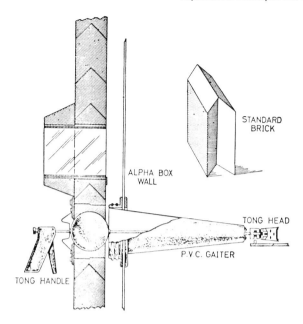

STANDARD
BRICK

ALPHA BOX
WALL

TONG HEAD

P.V.C. GAITER

TONG HANDLE

Fig. 7.15.   Typical 4 in. lead wall assembly.   Courtesy:
UKAEA, Harwell, England.

To illustrate some of the ramifications of this process on equipment and
techniques we will take the rather complex case of whether the failure of
one fuel element in an LMFBR subassembly results in a "domino-effect". It
has been postulated that when a crack propagates through the cladding wall
of a ceramic-fuelled LMFBR fuel element that the accumulated fission gas will
rush out and form a bubble between adjacent elements, especially at wire or
grid spacers, resulting in local overheating and another failure, which
propagates to the next element and so on.

To test whether this happens, a large sodium loop (SLSF or Sodium Loop Safety
Facility) was designed to be irradiated in the Engineering Test Reactor (ETR).
The loop would pump sodium past a pre-irradiated cluster of fuel elements, one
of which might be modified to fail during a test. The loop was carefully
instrumented to detect the failure sequence, which could be triggered by a
flow coast-down process. The large loop, or rather series of loops, required
extensive design and testing in conjunction with the ETR operations staff.
A special transport cask was built to carry the loops from HFEH/N, where they
are loaded with pre-irradiated fuel pins, to ETR on a large truck. HFEF/N
was modified to accommodate this cask and the loops (Fig. 7.11). The
detailed post-test analysis of the loop involves making many diametral cuts
along the fuelled section, examining each cut at magnifications between
50× and 1000× and then using more sensitive techniques to interpret the
microstructures.

Several tests have been run, at least one of which showed unusual material behaviour requiring very detailed metallurgical studies at the AGHCF in order to determine what had happened. This required the determination of very localized metallurgical reactions using an energy-dispersive X-ray analyser on the Scanning Electron Microscope that can analyse compositions within a 10 μm diameter spot. It was necessary for the scientist responsible for this reactor safety programme and the metallurgist performing the detailed analyses to maintain a very close dialogue so that the test could be properly interpreted and the next tests designed in the optimum manner.

A different example is the failure of LWR fuel elements through a PCI process. Fuel element failure statistics were built up from both reactor operations (e.g. the "sipping" process which allows early detection of failures) and poolside NDE tests. It became apparent that elements were failing somewhat earlier than predicted. Selected elements were sent to the Battelle Columbus hot cell — a large concrete cell designed for the examination of LWR fuel elements. The failure regions in the Zircaloy cladding were so small that only a few were detected and sectioned for examination. In the meantime suspicions grew that PCI might be due to stress corrosion cracking induced by one or more volatile fission products. Accordingly sections of cladding from irradiated LWR fuel elements were examined by SEM and Auger techniques, suspicion becoming focused on iodine. Unfailed sections of irradiated tube were pressure tested at temperature in the AGHCF in the presence of iodine vapour until they failed. The results showed more rapid failure in the presence of iodine.

In the meantime General Electric Company developed metal coatings on the inner wall of the cladding as a chemical and mechanical barrier against PCI. These were tested in the GETR at Vallecitos and gave improved performance. Full size subassemblies are now under test in commercial reactors. The moral of this story is that one needs very large and sophisticated resources to engage and compete effectively in the business of fuel element development and sales.

REFERENCES

1. Walton, G. N. (ed.), *Glove Boxes and Shielded Cells for Handling Radioactive Materials*, Academic Press/Butterworths Scientific Publications (1958).
2. *Proceedings of Conference on Irradiation Experimentation in Fast Reactors*, Jackson Lake Lodge, Wyoming (1973), American Nuclear Society (1973).
3. Fast Flux Test Facility Irradiation Services (August 1977). Obtainable from Director, FFTF Project Office, P.O. Box 550, U.S. Department of Energy, Richland, WA 99352.
4. Gillette, J. L., *et al.*, Argonne National Laboratory Report ANL-76-78. Technical Information Center (1977).

CHAPTER 8

Water Reactor Fuel Performance

The commercial power reactor field is dominated by three main types:
pressurized light-water reactors (PWRs), boiling light-water reactors (BWRs)
and heavy-water reactors (HWRs). The latter may be subdivided into boiling
and nonboiling types; the later CANDU systems and the British SGHWR are
boiling HWRs. All of these systems use UO_2 as the fuel and Zircaloy-2 or -4
as the cladding. Hence, there are a number of common features in their fuel
element performance.

The performance record of water reactors worldwide is impressive [1]. Over
9 million fuel rods have been irradiated to 1979, including 3.5 million HWR
rods. Although past experience has been mixed, the failure rates of the
current generation of fuel rods are at an acceptably low level, i.e. they
permit a high plant availability [2].

PWR

In all PWRs the fuel rods are assembled into a square array held together by
spring clips or spacer grids and by nozzles at the top and bottom (Fig. 8.1).
The assembly includes a number of control rod guide tubes in which approximately 16 control rods slide; these rods are connected to a common actuator
rod at the top of the assembly. This is known as the rod cluster control
(RCC). The PWR assembly is open at the sides, allowing cross-flow of the
coolant, independent expansion of the structural components and the fuel rods,
and easy visual inspection before and after discharge (Fig. 8.2).

The individual fuel rods vary in diameter from 9.63 mm (0.379 in) to 11.8 mm
(0.440 in) and are between 3.71 m (146 in) and 4.09 m (161 in) in length.
The active fuel length is shorter by about 230 mm, or 11 in, which represents
the plenum length. The Zircaloy-4 cladding thickness is about 0.6 mm (25
mil). The UO_2 fuel pellets have a density of between 94% and 95% of the
theoretical value [2].

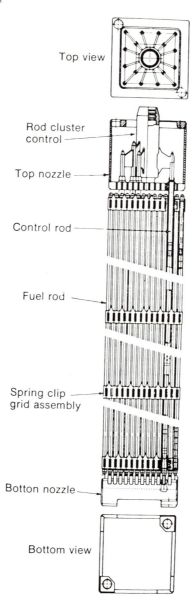

Top view

Rod cluster
control

Top nozzle

Control rod

Fuel rod

Spring clip
grid assembly

Botton nozzle

Bottom view

Fig. 8.1.   PWR fuel assembly.   Source: WASH-1250.

Fig. 8.2. PWR fuel assemblies. Courtesy: Electric Power
          Research Institute.

There are four PWR fuel vendors in the USA and at least four in Europe and
Japan. The earlier fuel assemblies were characterised by a smaller number of
larger diameter rods, operating at higher linear heat generation rates (LHGR).
To reduce the incidence of failures and to meet more stringent licensing
standards the trend has been to reduce rod diameters and hence LHGRs.
Table 8.1 illustrates this trend, while Fig. 8.3 shows the number and type of
PWR fuel rods in operation up to 1977.

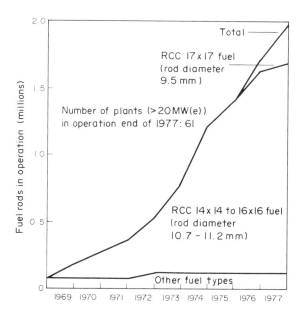

Fig. 8.3. Number of PWR fuel rods in operation. Source:
          R. Garzarolli, *et al*.

TABLE 8.1. Main Fuel Types for Water Reactors (Western countries)

Reactor		Former fuel design				Present fuel design			
Type	Manufacturer	Number of rods	Rod o.d. (mm)	Ave. LHGR (W/cm)	Number of plants[a]	Number of rods	Rod o.d. (mm)	Ave. LHGR (W/cm)	Number of plants[a]
PWR	Babcock & Wilcox and BBR	15x15-17	10.92	177-200	8	17x17-25	9.63	170-180	0
	Combustion Engineering	14x14-4x5	11.18	184-194	6	16x16-4x5	9.70	170-180	0
	Kraftwerk Union	14x14-16 / 15x15-20	10.75	177-226	1 / 3	16x16-20	10.75	207	2
	Westinghouse[b] and Framatome	14x14-17 / 15x15-21	10.72	150-220	17 / 16	17x17-25	9.50	170-180	6
BWR	ASEA Atom	8x8	12.25	126-155	3	8x8-1	12.25	158-175	2
	Gen. Electric[c] and Kraftwerk Union	6x6[d] / 7x7[d]	14.3	150-234	7 / 30	8x8-2	12.5	170-200	4
HWR	AECL[e] (D$_2$O-CANDU)	19/28	15.2	80-260	8	37	13.08	210	2
	Kraftwerk Union (PHWR)	37/36	11.9	116/232	2	-	-	-	-
	UKAEA/BNFL (SGHWR)	36	16.0	210	1	60	12.2	-	0

[a] Number of plants (>20 MW(e)) in operation since January 1978.
[b] Including Mitsubishi.
[c] Including Hitachi and Toshiba.
[d] 8x8 reload since 1973/74.
[e] 50 cm length of bundle.

Source: F. Garzarolli, R. von Jan, H. Stehle, *Atomic Energy Review*, __17__, 1 (1979).

BWR

The BWR fuel assemblies differ from the PWR in a number of respects. Most importantly, the control rods are external to the fuel assemblies; they are + shaped plates that move in the space between four fuel assemblies (Fig. 8.4). The fuel assemblies therefore contain only fuel rods, arranged in earlier designs in 6 × 6 or 7 × 7 arrays, but now uniformly in 8 × 8 arrays. These rods are spaced by Zircaloy grids. There are eight tie rods which hold the top and bottom of the assembly together. The assembly is contained within a Zircaloy box or channel which confines the coolant flow, i.e. there is no cross-flow. BWR fuel rods are clad in Zircaloy-2. The rod diameter and cladding thickness are somewhat larger than in PWRs; ∿12.5 mm (0.493 in) o.d. and 0.864 mm (34 mil) thickness. The fuel column length and plenum size are similar to that in PWRs.

General Electric is the only BWR vendor in the USA, but Exxon and (recently) Westinghouse make reload fuel assemblies. ASEA-Atom and KWU make BWR fuel in Europe.

The performance characteristics of BWR fuel are shown in Table 8.1; again there is an historical trend towards smaller rod diameters and lower LHGRs, resulting in the standardization of the 8 × 8 assembly. The number of the different types of BWR fuel rods in operation up to 1977 is shown in Fig. 8.5.

In comparing PWR and BWR fuel rod performance it should be noted that the PWR coolant pressure is roughly twice that of the BWR. The residence times in the cores to the target burn-ups are: 3 years in the PWR to achieve a mean maximum burn-up of 28-34 GWd/t(U) and 4 years in the BWR to achieve 22-28 GWd/t(U).

HWR

The CANDU Heavy Water Reactor fuel assembly design is quite different from that for PWRs and BWRs. The fuel is natural enrichment UO_2 in pellet form, clad in Zircaloy-2 tubes. Because the fuel is not enriched, it must be loaded and unloaded on-power. The coolant tubes are arranged horizontally to make this operation easier and the fuel assemblies are fairly short (50 cm or ∿20 in) in order to achieve a higher mean burn-up and because the core length is so great (∿6 m or 23 ft). The UO_2 pellets are sealed into 1.31 cm (0.515 in) diameter rods with 0.4 mm (16 mil) wall thickness. Note that the wall thickness is only half that of the smaller diameter LWR rods. Each rod has bearing pads welded to its o.d. in a number of places to separate it from its neighbours. Thirty-seven rods are bundled together and welded to two end plates (Fig. 8.6). The fuel residence time in the core is ∿470 full power days to reach a design mean maximum burn-up of 7 GWd/t(U). This type of assembly is made by AECL and by KWU in Germany.

FAILURE

Failure of a fuel rod is defined as the appearance of an opening or openings across the cladding wall (or end welds) sufficient to permit egress of fission products and/or fuel; and/or ingress of coolant. The significance of a failure is in its effect on reactor operations, maintenance and district safety. As one might expect, failures usually start small and grow. Typically a pinhole appears in the cladding and eventually may grow into a crack; this may be aided by the ingress of water during a shutdown, which can flash to steam and add to internal pressure on reactor start-up.

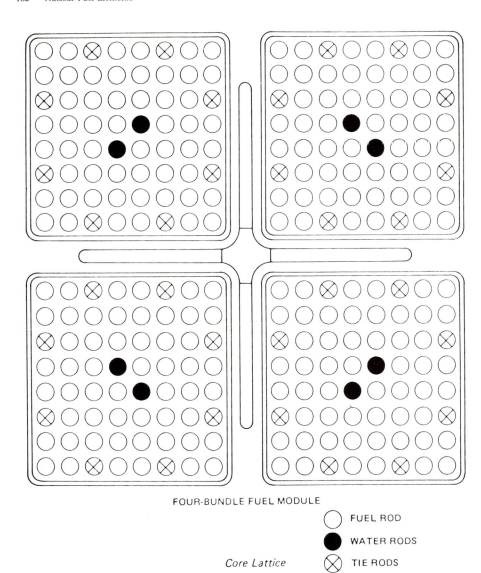

FOUR-BUNDLE FUEL MODULE

Core Lattice

◯ FUEL ROD

⬤ WATER RODS

⊗ TIE RODS

Fig. 8.4.   Boiling water reactor core components.
<u>Courtesy</u>: General Electric Company.

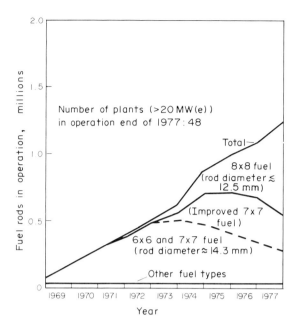

Fig. 8.5. Number of BWR fuel rods in operation.
Source: R. Garzarolli *et al.*

Failures can occur at different times in the fuel rod lifetime, usually described by a "bathtub"-shaped curve (Fig. 1.6). Poor quality assurance can lead to "infant mortality", e.g. through weld failure. The failure rate versus burn-up then remains low until a certain burn-up level when fuel swelling and fission gas pressure strains the cladding beyond its design level. It will emerge in later discussion that this simple picture may be distorted by unforeseen phenomena and by the mode of reactor operation.

In the case of an early-in-life failure, little will happen at failure; subsequent irradiation will generate fission products that will escape into the coolant [3]. In later failures fission gases will escape rapidly and failure can be detected almost at once. In BWRs the fission gases are removed from the circulating water by the steam separators, pass through the turbine to the condenser and the air ejector which extracts these fission gases which are diverted to a hold-up line, where they decay, and finally are absorbed on a charcoal bed for permanent storage and decay.

In PWRs the fission gases are mainly confined to the primary circuit, which is fairly leak-tight. Some gas may leak through the steam generators — this is separated onto charcoal beds. Iodine isotopes (including the xenon precursors) tend to stay in solution in the coolant water and can lead to radiation build-up in the coolant circuits.

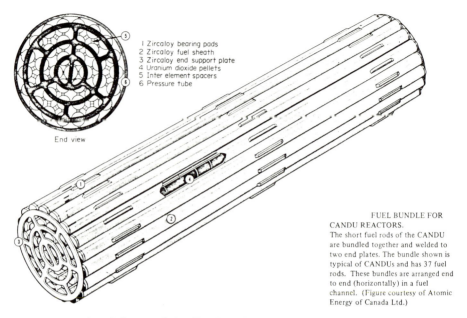

1 Zircaloy bearing pads
2 Zircaloy fuel sheath
3 Zircaloy end support plate
4 Uranium dioxide pellets
5 Inter element spacers
6 Pressure tube

End view

FUEL BUNDLE FOR
CANDU REACTORS.
The short fuel rods of the CANDU
are bundled together and welded to
two end plates. The bundle shown is
typical of CANDUs and has 37 fuel
rods. These bundles are arranged end
to end (horizontally) in a fuel
channel. (Figure courtesy of Atomic
Energy of Canada Ltd.)

Fig. 8.6.   Fuel bundle for CANDU reactors.   Source:
Atomic Energy of Canada Ltd.

Generally a reactor core containing 50,000-60,000 fuel rods can tolerate up
to 5-10 "leakers" without shutting down or reducing power. This is a
significant figure, because the whole aim of commercial nuclear power
generation is to keep operating at full power; unscheduled shutdowns or even
power reductions are very expensive to the utility and also to the fuel
supplier if he has contracted for a guaranteed performance (burn-up). The
industry has not yet reached this goal of a 0.01-0.02% failure rate, but it
is getting close. Early fuel showed high failure rates, PWRs being ∿0.1%
and BWRs 0.76%, so that very substantial progress has been made [4] and there
is a consensus that only one significant cause of fuel rod failure remains:
pellet-cladding interactions, which we will discuss shortly.

FAILURE MECHANISMS (see Table 8.2)

Hydriding of Zircaloy [5]

Zircaloy has a high affinity for both oxygen and hydrogen, hence it is to be
expected that the water coolant will provide both elements to the fuel
cladding. Normally Zircaloys react slowly with water to give a modest
surface oxide film. Some hydrogen enters into solution and diffuses away
from the surface. Local enhancement of the hydrogen pickup, to an extent
that exceeds the rate of diffusion into the interior, leads to the formation
of zirconium hydride. This is a lower density phase and it will therefore
lead to local volume changes resulting in blistering or to cracking. If this
occurs on the inside wall of the cladding it often has a "sunburst"
appearance (Fig. 4.6). These sunbursts with their associated cracks can lead
to local penetration of the cladding, as the figure shows.

TABLE 8.2.   Main Factors that have Limited Fuel
Performance and Remedies

Factors	PWR	BWR	Remedies
Hydriding of Zr	X	X	Elimination of moisture in fabrication
Scale deposition		X	Elimination of copper tubing from feedwater heaters
Enrichment errors	X	X	100% rod scan
Clad collapse	X		Prepressurized cladding, stable pellets
Pellet densification	X	X	Stable structure
Other manufacturing and handling defects	X	X	<5% of defects
Clad corrosion or fretting		X	Rare; control of clad quality and cleaning; spacers
Clad growth and bowing	X		Tubing tolerances; axial clearances; spacer design
Channel bulging		X	Thicker wall; control of residual stress
Pellet-clad interaction on power increases	X	X	1. Slow power rise (PWR) 2. Plus local power shape control (BWR) 3. Plus fuel design changes (both)

Source: M. Levenson and E. Zebroski, *Ann. Rev. Energy*, 1, 101 (1976).

The source of water (and hydrogen) on the inside of the cladding can only be from the fuel. As described in Chapter 2, UO_2 is fabricated by a pressing and sintering process, the product of which contains small pores. If the fuel density is less than ∿93% theoretical, i.e. it has more than ∿7% porosity, the pores may be connected to one another and to the surface with a measured area on the order of a few m^2/g. Water vapour up to 50 ppm can be adsorbed on these pores and released when the fuel undergoes fission heating. This water vapour release led to hydride failures, especially in BWRs, but also in PWRs but not in HWRs. This susceptibility in BWR fuel rods may have been due, in part, to the presence of fluorine in the fuel (from prior processing) which aided in the breakdown of the surface oxide film to form the sunburst. Hydride failures can be largely eliminated by the use of very dry atmospheres during post-sintering operations or by a vacuum degassing process; these bring the residual moisture down to an acceptable level of 1 ppm or less. One can also incorporate a hydrogen getter in the plenum. Care must also be taken to exclude all other hydrogenous materials from the fuel rod components during manufacture, since these can also contribute to hydriding. By taking these precautions the failure incidence due to hydriding has dropped to a low level.

Rod Bowing and Zircaloy Growth

Zirconium or Zircaloy, having a hexagonal crystal structure, is subject to anisotropic growth under irradiation [6]. All fabricated Zircaloy tubing has a texture, i.e. the grains are not randomly oriented, so that there is a net growth effect in Zircaloy fuel rods or assemblies. Furthermore, the growth may be different in different fluxes and temperatures in the reactor leading to differential growth effects and to bowing. Bowing may also be

caused by nonuniform stress effects resulting from radiation-induced stress
relaxation or from interactions between the fuel rods and the spacers [7].
These effects have not caused significant incidence of failures in the past,
but the push towards higher burn-up targets, and hence higher neutron
fluences on the cladding and ducts, may result in more serious bowing and/or
growth problems.

Fretting and Wear

Flow-induced vibrations can cause fretting and wear between the cladding and
the grids. The open structure of the PWR assembly makes this sytem more
susceptible. Some failures have occurred because of the hold-down springs
being too weak to stop axial oscillations. In BWRs vibration has caused
fretting of the flow channels. In general, redesign and more attention to
quality control have eliminated fretting and wear failures.

Pellet Densification

This was discussed earlier: to recapitulate, under irradiation the fine pores
($\leqslant 1$ μm diameter) in UO_2 disappear, leading to densification of the pellets[8].
While the local effect is small ($\sim 1\%$ increase in density) the cumulative
effect in a 12-ft fuel column is large enough to cause the column to shrink
by a few inches relative to the cladding. This left an empty region which
collapsed under the PWR coolant pressure (Figs. 8.7 and 8.8). Only a small
fraction of the collapsed tubes developed leaks, but the phenomenon caused
concern on such issues as power spiking and neutron streaming. Two remedies
were found for this effect. First, all PWR rods were prepressurized to
~ 1000 psi of helium to prevent cladding collapse. Second, fuel manufacturers
changed their fuel fabrication processes to eliminate most of the small pores
from the final microstructure, generally by raising the sintering temperature
(Fig. 8.9) [9].

Pellet-Cladding Interaction (PCI) [10]

The most persistent and troublesome cause of cladding failure is labelled
Pellet-Cladding Interaction or PCI. We can discuss PCI from both the
empirical and the mechanistic standpoints.

Empirically it has been observed on numerous occasions that fuel rod failures
were the consequence of a certain pattern of reactor operations or manoeuvres.
These may be characterized by Fig. 8.10. The fuel assembly or fuel rod is
irradiated for a finite time to a given burn-up BU and is operating at a
power lower than the maximum (it may even be zero, i.e. shutdown). The power
is then raised at a ramp rate dP/dt through a ramp height ΔP to a level of
power P, where it dwells for time T before it fails. For failure to occur
all five parameters (BU, dP/dT, ΔP, P and T) must all be in a critical range.
Critical values of ΔP vary with BU (Fig. 8.11).

Attemps have been made to model PCI incidence in terms of these five parameters
AECL workers developed the FUELOGRAM model to predict PCI defect probabilities
for CANDU fuel bundles [11]. This was modified to extend the predictions to
PCI in LWR fuel rods (PCI-OGRAM). In a parallel study workers at PNL
developed the PROFIT model to analyse PCI behaviour. In the latter an attempt
was made to attach mechanistic correlations to the four operating parameters:

1. P, the post-transient power, is related to the temperature for relatively prompt release of fission products from the fuel.
2. ΔP, the transient power increase, is related to the differential expansion of the fuel and cladding.
3. The burn-up, BU, determines the fission product inventory within the fuel which affects the amount of potentially corrosive fission products in the fuel, the amount of fuel swelling and the fission gas pressure. It also determines the neutron fluence (damage) in the Zircaloy cladding which affects the amount of creep and stress relaxation that has occurred — principally in terms of cladding collapse onto the fuel pellets.
4. The dwell time, T, is postulated to be the time to failure under a particular set of chemical and mechanical conditions.

There is a hint above to the likely mechanism(s) of PCI. Remember that the cladding is subject to a high external pressure from the coolant which forces it down onto the pellet at all times. Thus, high local stresses are induced in the cladding at pellet-pellet interfaces and where thermal-shock cracks reach the pellet surface. During a power ramp the fuel expands faster than the cladding, increasing the local stresses. At the post-transient power level, volatile fission products, including iodine and caesium, are released into the fuel-clad gap and cause stress-corrosion cracking (SSC) of the Zircaloy in the regions of high local stresses. The dwell time, T, is then the rupture time under the particular SCC conditions [12,13,14].

The problem has been tackled on several levels. First, fuel manufacturers have had to specify to the reactor operators certain rules or conditions for reactor operations to minimize or avoid PCI failures. This primarily limits the rate of power increases, but also influences local power shaping. This is annoying to the operators because it results in a significant net loss in power output. In a typical example, the Vermont Yankee BWR core [15] is raised to 45% of rated power fairly rapidly, then power is raised at 7 MW/hr until the upper limit of flow control is reached. If full rated power is not reached, a hold period occurs, after which flow is reduced, control rods are slowly withdrawn and the process repeated until full power is reached; this usually takes three iterations. In the years 1975 and 1976, over 4% of the capacity factor was lost, i.e. 2% per year.

Lost capacity is being reduced in a number of ways. First, the change to smaller rod diameters has led to reduced fuel temperatures and a lower tendency to release fission products. Prepressurization, including on BWR fuel rods, has helped to reduce cladding stresses. Better management of fuel and of power shaping has also helped.

On another level much effort has gone into understanding the PCI mechanism. While the iodine stress corrosion cracking mechanism is strongly supported [16], there are anomalies that cannot yet be explained, e.g. the benefits or mechanisms of prepressurization are not clear, and work continues to understand the process completely. In the meantime semi-empirical remedies are being tested. The most advanced of these are the use of coatings or liners between the fuel and the cladding. The only one that is in commercial use is the AECL's CANLUB design in which a graphite coating is placed between fuel and cladding. This probably reduces fuel-cladding mechanical interactions and acts as a diffusion barrier to fission products. GE has published results of tests on BWR fuel rods with a copper coating on the inner cladding wall and with a zirconium liner in the fuel-clad gap. Both have given promising performances.

Fig. 8.7. Example of cladding collapse from Beznau I PWR.
Source: G. Roberts *et al.*, Central Electric
Generating Board, England.

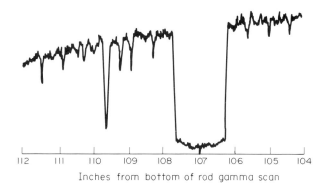

Inches from bottom of rod gamma scan

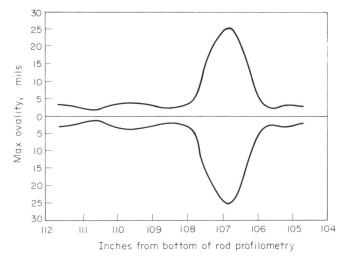

Inches from bottom of rod profilometry

Fig. 8.8. Section of Beznau I Region 2 fuel rod with a
short gap in the fuel column. <u>Source</u>: ibid.

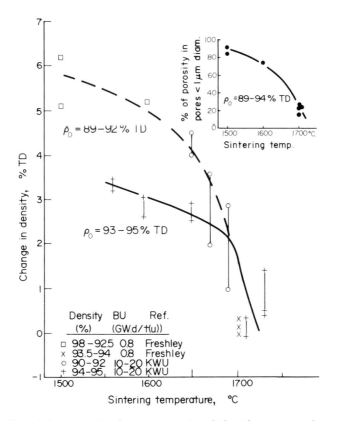

Fig. 8.9.   Correlation between densification propensity
and sintering temperature.

EPRI has supported a programme to test prepressurized fuel rods in which annular UO_2 pellets are doped with Nb_2O_5 to give a large initial grain size, which should reduce the fission product release, and may increase the fuel plasticity and so reduce the net amount of fuel cracking. The annular pellet design will reduce the fuel centre temperature and provide voidage to accommodate fuel swelling and relieve cladding stresses.

The efforts being expended by the nuclear industry at this time are such that one can expect the restrictions on reactor operations to be steadily decreased with a consequent increase in total power output.

LICENSING AND REGULATION

Even before the Three Mile Island incident NRC regulations controlling thermal reactor fuel qualifications for the licensing of cores were becoming increasingly stringent. It is likely that this trend will accelerate. Strasser and Lindquist [17] have quoted an example of a 3000 MWt PWR where the original design operating limit was 18.4 kW/ft. Acceptance criteria for the emergency core cooling system (ECCS) and a new fuel densification model reduced this to 13.2 kW/ft. Since the fuel and core design were unaltered, a margin between this limit and actual operating conditions was maintained by placing maximum reliance on chemical shim for reactivity control and reducing the use of control rods.

A new fuel can be introduced into a power reactor in one of two ways: either as an initial full core or regionally as a reload batch [2]. Smaller scale tests may be performed with lead test assemblies (LTAs), but care has to be taken to ensure that these are tested under typical conditions (because of local variations across a core). In general, PWR fuel manufacturers have chosen the full core reload route, while BWR fuel manufacturers have chosen the partial reload route. Exxon occupies a unique position in being a supplier of BWR and PWR fuel assemblies but not a reactor manufacturer. Hence their fuel is always introduced in reload batches. To illustrate what the NRC requires for the verification of the performance of a new design of fuel assembly the following quotation is given from the NRC public records:

17X17 FUEL SURVEILLANCE PROGRAM

To provide verification of the reliable performance of the 17x17 fuel assemblies, a supplemental fuel surveillance program will be conducted at Plant A and B. The program will consist of a visual inspection of all the peripheral rods in the initial core fuel assemblies as they are discharged into the spent fuel pool. Approximately one-third of the initial core fuel assemblies will be inspected during each of the first three refueling periods. The visual inspection will include observations for cladding defects, fretting, rod bowing, corrosion, crud deposition and geometric distortion.

The initial core loading will contain two precharacterized fuel assemblies in each of the three fuel zones of the core. Precharacterization will establish baseline data that could be used to facilitate the evaluation of fuel performance, dimensional changes, or any anomalies that might be evident during the visual examination. If any anomalies are detected during the visual examination, further investigation will be performed. Depending on the nature of the observed condition, this further examination could include appropriate surface, dimensional, or gamma inspections of the fuel assemblies. If the fuel assembly design enables reconstitution, individual fuel rods may also be examined. Under unusual circumstances, destructive examination of a fuel assembly or individual fuel rods may be required, but this would not be accomplished on site or within the time of the refueling outage. The NRC will be advised of the normal refueling schedule, and will be notified at least ten days in advance of any planned supplemental fuel surveillance inspections so that it may observe the inspections and the resultant evaluations of the fuel assembly performance. Following the inspection of a significant number of assemblies and prior to reactor re-startup, an oral report of the results will be made to NRC. Within 30 days of completion of the total inspection, a written report will be submitted to NRC.

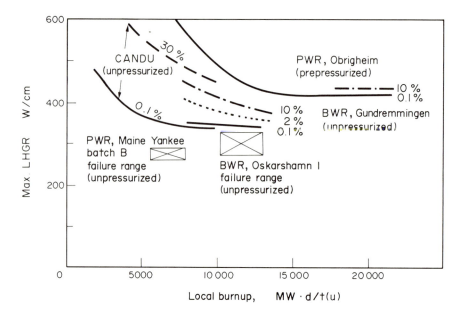

Fig. 8.10.   Correlation between maximum power during a
fast ramp and the defect rate by PCI of fuel
rods of various designs.
Source: Garzarolli *et al*.

The detailed programme above has been discussed with the utilities and PWR
vendors. Definite commitments for this programme have been secured from
Westinghouse on two of the following plants (Trojan, Beaver Valley, Salem 1,
Farley) with 17 × 17 fuel and Combustion Engineering for one plant
(Arkansas 2) with 16 × 16 arrays. The next C-E 16 × 16 plant (San Onofre 2)
is scheduled for late 1980. Plants that use the Babcock & Wilcox 17 × 17
design are not scheduled to operate until 1982.

Note the emphasis on precharacterization, i.e. thorough physical, chemical
and mechanical measurements of the fuel, cladding and assembly before (and
after) irradiation. As an overlay on these requirements the NRC places
certain basic conditions on fuel rod performance, e.g. the fission gas
pressure within the fuel rod must not exceed the coolant pressure.

The Three Mile Island incident showed that a loss of coolant accident (LOCA)
could happen, although it was not the "classical" LOCA accident which the
NRC and its overseas counterparts had been considering. It is not the aim
of this book to discuss TMI, especially at this time when the core has yet
to be examined. Suffice it to say that the TMI incident was a small leak
type of accident (the open pressure relief valve) which was improperly
handled by the operations staff. In particular they turned off the ECCS to
avoid damage to their pumps and in so doing grossly overheated (or undercooled)
the core.

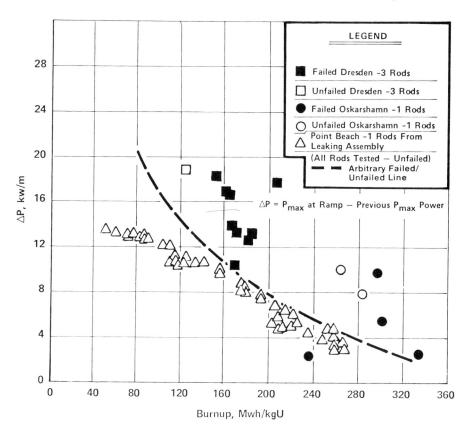

Fig. 8.11.   Relationship between power change ΔP and
burn-up for failure-no failure condition.
Courtesy: Nuclear Regulatory Commission.

The "classic" LOCA scenario is the loss of coolant through a large rupture
in the primary pressure boundary (probably a pipe) [18]. The core is shut
down by insertion of the control rods and loss of moderator, but the fuel
continues to generate fission product decay heat (initially at ~10% of the
full power but decaying exponentially, Fig. 8.12). At this stage the fuel
rods are cooled, albeit inadequately, by the boiling steam-water mixture.
During this "blowdown" the ECCS comes into operation, pouring water onto the
core from the top and/or bottom: Fig. 8.13 illustrates schematically the
arrangement in a typical PWR system.

The NRC requires that the design be such that a "coolable geometry" be
maintained at all times, so that the core cannot further overheat and enter
the meltdown phase. Much work has been done and continues to be done to
define the limiting conditions for maintaining a coolable geometry.
Calculations, model tests and full-scale tests in the LOFT (Loss of Flow Test)
facility in Idaho have been carried out to define the cladding temperature
history and behaviour in a "typical" LOCA. The NRC has assumed that these
tests and calculations contain uncertainties and has, therefore, adopted a

very conservative attitude in defining the limits for a coolable geometry; this is clearly shown in Fig. 8.14 which plots the postulated maximum cladding temperature versus time from the pressure boundary rupture.

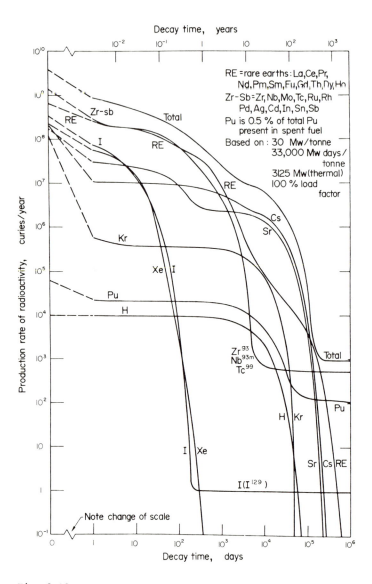

Fig. 8.12. Decay of radioactivity in thermal reactor fuel. Source: Anthony V. Nero, Jr., A Guidebook to Nuclear Reactors, University of California Press, 1979.

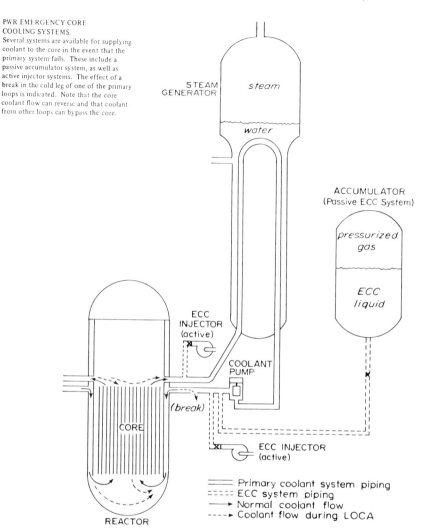

PWR EMERGENCY CORE
COOLING SYSTEMS.
Several systems are available for supplying
coolant to the core in the event that the
primary system fails. These include a
passive accumulator system, as well as
active injector systems. The effect of a
break in the cold leg of one of the primary
loops is indicated. Note that the core
coolant flow can reverse and that coolant
from other loops can bypass the core.

STEAM
GENERATOR   *steam*

water

ACCUMULATOR
(Passive ECC System)

pressurized gas

ECC liquid

ECC
INJECTOR
(active)

COOLANT
PUMP

(break)

CORE

ECC INJECTOR
(active)

══════ Primary coolant system piping
╌╌╌╌╌ ECC system piping
────► Normal coolant flow
----► Coolant flow during LOCA

REACTOR

Fig. 8.13.   PWR emergency core cooling systems.
             <u>Source</u>: ibid.

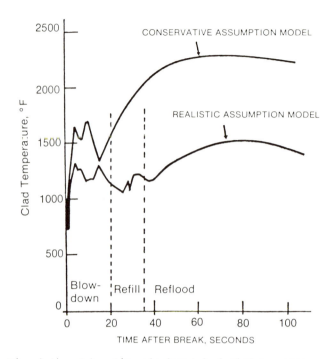

Fig. 8.14.   Schematic calculated fuel clad temperatures
for a PWR LOCA.   Source: ibid and WASH–1250.

Before stating the current acceptance criteria for emergency core cooling
systems, let us consider what the cladding has to sustain during the sequence
described above. As the coolant flow stops the cladding heats up (as shown
in Fig. 8.14) and oxidizes. The fuel also heats up and releases fission
gases which increase the internal pressure to a stage where the hot (and
hence weaker) cladding deforms. This usually produces ballooning, i.e. gross
deformation without rupture (Fig. 8.15). The ECCS then rapidly cools the
cladding, which may fail by thermal shock, e.g. shatter into pieces. If it
survives, the question arises: can the fuel rods stand up to vibration and
jolting while they are unloaded from the reactor and placed in a storage
area or a transfer cask?

An understanding of the cladding behaviour under these conditions has proven
to be a complex metallurgical and engineering study, yet one that is very
important from both a licensing standpoint and a reactor operations standpoint. The latter arises because there is so much conservatism built into
the cladding temperature calculations that reactor powers are restricted,
resulting in loss of power output. Hence, the work must continue in order to
reduce the level of uncertainty or conservatism.

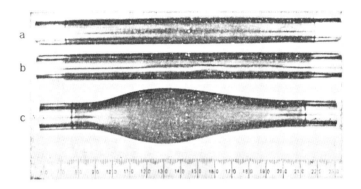

Fig. 8.15.   Specimens from SGHWR fuel cans after LOCA
tests. Source: D. O. Pickman, Physical
Metallurgy of Reactor Fuel Elements, The
Metals Society, 1975.

The current NRC regulations date from 28 December 1973, when a document was
published called: "Acceptance Criteria for Emergency Core Cooling Systems for
Light Water Reactors", Docket No. RM-50-1. The criteria are:

1. *Peak cladding temperature:* The calculated maximum fuel-element cladding
   temperature shall not exceed $1477^{\circ}K$ ($1204^{\circ}C$).
2. *Maximum cladding oxidation:* The calculated total oxidation of the cladding
   shall nowhere exceed 0.17 times the total cladding thickness before
   oxidation. (A more detailed discussion of the oxidation criterion
   follows.)
3. *Maximum hydrogen generation:* The calculated total amount of hydrogen
   generated from the chemical reaction of the cladding with water or steam
   shall not exceed 0.01 times the hypothetical amount that would be
   generated if all the metal in the cladding cylinders surrounding the fuel,
   excluding the cladding surrounding the plenum, were to react.
4. *Coolable geometry:* Calculated changes in the core geometry shall be such
   that the core remains amenable to cooling.
5. *Long-term cooling:* After any calculated successful initial operation of
   the ECCS, the calculated core temperature shall be maintained at an
   acceptably low value and decay heat removed for the extended period of
   time required by the long-lived radioactivity remaining in the core.

The basis for these criteria were a number of experiments on undeformed
cladding heated by various means and then rapidly cooled. The limiting
conditions were defined as those below which the cladding remained intact.

Since that time a great deal of additional work has been performed in which a much better understanding has been obtained of the properties of highly oxidized Zircaloy. Some new criteria that have been recommended (with much data to support them) are [19]:

1. *Capability to withstand thermal shock during LOCA reflood.* The calculated thickness of the cladding with ≤0.9 wt % oxygen, based on the average wall thickness at any axial location, shall be greater than 0.1 mm (4 mil).
2. *Capability for fuel handling, transport and interim storage of oxidized fuel assemblies.* The calculated thickness of the cladding with ≤0.7 wt % oxygen, based on the average wall thickness at any axial location, shall be greater than 0.3 mm (12 mil).

These criteria incorporate within them consideration of the peak cladding temperature and maximum hydrogen generation. By comparison, the 1477°K and 17% oxidation limits are quite conservative. This can be seen in Fig. 8.16 which maps out the failed and unfailed regions as a function of cladding temperature and amount of cladding reacted (oxidized). The 17% limit is plotted on this figure: it is well below the experimental failure regime.

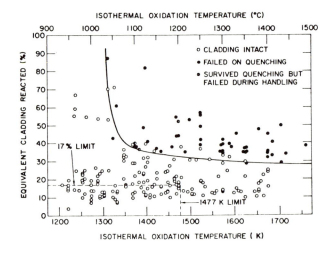

Fig. 8.16. Failure map for Zircaloy-4 cladding by thermal shock or normal handling relative to the equivalent-cladding-reacted parameter and maximum isothermal oxidation temperature after rupture in steam. Source: H. Chung and T. Kassner, Argonne National Laboratory.

To test cladding for embrittlement in relation to these criteria it is suggested that a measurement by optical microscopy be made of the α- and β-phase layers in cross-sections of Zircaloy cladding and that impact measurements be made on ∿200 mm (∿8 in) lengths of cladding at a temperature ≤400°K (127°C). The interested reader is referred to Chung and Kassner's [19] report for details.

Although similar criteria may be applied to HWRs, it has to be remembered that these reactors confine their fuel within pressure tubes which are surrounded by the D_2O moderator. Hence, if a LOCA occurs it will be very localized and the D_2O will help to cool that region. The worst that might happen is that some fuel and pressure tubes would be damaged, but a meltdown could not occur.

REFERENCES

1. Garzarolli, F., von Jan, R. and Stehle, H., *Atomic Energy Review*, 17, 1 31 (1979).
2. Houston, M. Dean, U.S. Nuclear Regulatory Commission Reports NUREG-0633 (December 1979) and NUREG/CR-818 (January 1981).
3. Zebroski, E. and Levenson, M., *Ann. Review of Energy*, 1, 101 (1976).
4. Proebstle, R. A., *et al.*, *Commercial Nuclear Fuel Technology Today*, Toronto, p. 2 (1975).
5. Pickman, D. O., *Nucl. Eng. Des.* 33, 141 (1975).
6. Murgatroyd, R. A. and Rogerson, A., *J. Nucl. Mat.* 79, 302-311 (1979).
7. Montgomery, M. H., Mayer, J. T. and French, D., *Proceedings of Topical Meeting on Water Reactor Fuel Performance*, Am. Nucl. Soc., St. Charles, IL, p. 71 (1977).
8. Roberts, G., Cordall, D., Jones, K. W., Cornell, R. M. and Waddington, J. S. ibid, p. 92.
9. Freshley, M. D., *et al.*, J. Nucl. Mat. 62, 138 (1976).
10. Discussed at length in *Proceedings of Topical Meeting on Water Reactor Fuel Performance*, Am. Nucl. Soc., St. Charles, IL (1977).
11. Mohr, C. L., Pankaski, P. J., Hester, P. G. and Wood, J. C., Nuclear Regulatory Commission Report NUREG/CR-1163 and Pacific Northwest Laboratory Report PNL-2755, pub. NTIS (1979).
12. Garzarolli, F., *Kerntechnik*, 20, 27 (1978).
13. Roberts, J. T. A., *et al.*, Electric Power Research Institute Report NP-1024-SR (1979).
14. Pasupathi, V., *et al.*, Electric Power Research Institute Report NP-812 (1978).
15. Candon, J. D., *Proceedings of Topical Meeting on Water Reactor Fuel Performance*, Am. Nucl. Soc., St. Charles, IL (1977).
16. Yaggee, F. L., Mattas, R. F. and Neimark, L. A., EPRI Report NP-1155 (1979).
17. Strasser, A. and Lindquist, K. O., *Proceedings of Topical Meeting on Water Reactor Fuel Performance*, Am. Nucl. Soc., St. Charles, IL, p. 18 (1977).
18. WASH-1250, The Safety of Nuclear Power Plants (Light-Water Cooled) and Related Facilities, NTIS (July 1973).
19. Chung, H. M. and Kassner, T. F., Embrittlement Criteria for Zircaloy Fuel Cladding Applicable to Accident Situations in Light Water Reactors: Summary Report, ANL-79-48, NTIS (1980).

CHAPTER 9

Gas-cooled Reactor Fuel Elements

Commercial nuclear power has taken two distinct paths, each being characterized by a particular combination of coolant and moderator. In the previous chapter the water-cooled and moderated reactor fuel elements were discussed. These have been a North American development. In this chapter we will discuss the gas-cooled, graphite-moderated reactors that were developed in Europe. It should be noted, in passing, that water-cooled, graphite-moderated reactors were developed for plutonium production in the USA and for power production in the Soviet Union. However, these are not likely to seriously challenge either of the other types in world power production. Furthermore, information on their performance is less accessible, so we will not discuss them further.

The first electricity-producing gas-cooled reactors used CO_2 as coolant and were built in Britain and France with the additional purpose of producing plutonium for weapons. These evolved into larger-size commercial plants operated by the state utilities and versions were sold to Japan (Tokai) and to Italy (Latina). These reactors were relatively small (up to 300 MWe) and had relatively low coolant outlet (and hence steam) temperatures.

The next logical development was the Advanced Gas-cooled Reactor (AGR), the prototype of which was the Windscale AGR. This class of reactor also used CO_2 as its coolant, but the outlet temperature was raised to around 600°C, giving much higher thermal efficiency. Several large commercial AGRs have been built in Britain.

The final step in gas-cooled reactor development was the High-temperature Gas-cooled Reactor (HTGR). Developed concurrently in Europe and the USA, the HTGR system used helium as the coolant with an outlet temperature around 800°C, although this may be raised for applications other than steam turbine generators. The Dragon reactor in Britain and the Peach Bottom reactor in the USA were prototypes.

MAGNOX

The British Magnox reactors used (and still use) a uranium metal fuel at the natural enrichment level. The fuel needed to be stable under irradiation to at least 3000 MWd/te(U) and under thermal cycling. As mentioned earlier, modest alloying* to produce "adjusted" uranium led to good dimensional stability [1]. The cladding material was chosen to be Magnox AL80, an alloy of magnesium with 0.8 wt % aluminium and 0.002-0.050 wt % beryllium [2]. This alloy had good corrosion resistance up to 500°C in CO_2 and would not burn in air.

The uranium is in the form of cast and machined rods, about 2.75 cm (1.10 in) in diameter and ranging from 0.5 m (20 in) to 1.07 m (40 in) in length. The rods are heat-treated to produce a fine grain size which has a slight texture. Anti-ratchetting grooves are machined in the bars, which are then clad in finned Magnox cans (Fig. 9.1). The fins are necessary to produce the turbulence and extended surface area for adequate cooling of the fuel in the CO_2 coolant (Fig. 9.2). The α/β transformation in uranium occurs at 665°C. Cycling through this transformation leads to dimensional changes or growth with undesirable effects such as ratchetting and cracking. Hence the fuel temperature is kept below 665°C. The anti-ratchetting grooves lock the fuel to the cladding to reduce thermal cycling effects.

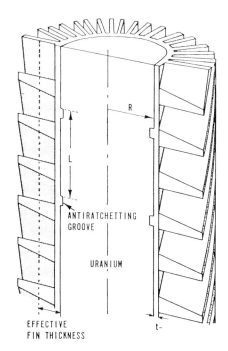

Fig. 9.1. Details of Magnox fuel element assembly.
Source: M. Simnad, Fuel Element Experience in Nuclear Power Reactors, 1970.

---

*∿400 ppm Fe and ∿1000 ppm Al.

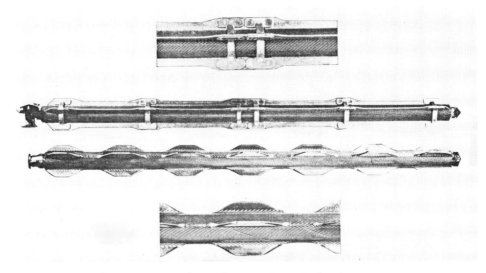

Fig. 9.2.   A helical-finned polyzonal (top) and a
herringbone fuel element (bottom).

The finned Magnox-clad fuel element is attached to a set of braces or
splitters — longitudinal sheets that hold the elements in their coolant
channels (Fig. 9.3). In later reactors (e.g. Hunterston) the elements are
placed inside graphite sleeves for loading in the reactor (Fig. 9.4). This
prevents fuel element bowing. There are variations in fuel element design
from reactor to reactor that are absent from the highly standardized water
reactor designs.

In general, the performance of these fuel elements has been very satisfactory
and has seen steady improvement [3]. Up to 1972 the CEGB had charged
1.1 million fuel elements in its reactors and had observed 1234 failures,
the causes being given in Table 9.1. The main cause of failure is attributed
to "body defects", most commonly caused by a brittle form of cracking between
large grains in the can wall. The problem appears to have been solved by
careful attention to the manufacturing process for the finned cans, i.e. by
avoiding those conditions that lead to a combination of texture and grain
growth. Fuel swelling has proved to be acceptably low in adjusted uranium,
giving an average discharge burn-up in excess of 4000 MWd/te(U) or ∿6500
MWd/te(U) maximum.

A form of fuel element failure in early elements was described as a "fast
burst". This occurred when a small leak formed in the end cap or end weld.
This was too small to give a signal on the burst cartridge (fuel rod)
detection gear (BCDG). However, CO_2 would enter the fuel element and oxidize
the uranium locally, forming a mound that would eventually burst open the
can and cause a large, sudden signal on the BCDG. This was overcome by
better inspection techniques before loading the fuel elements and more
sensitive leak detection methods in the reactor coolant channels.

A persistent problem in gas-cooled reactors is flow-induced vibration, i.e.
"rattling" of the fuel elements in the fast flowing gas stream that can lead
to damage to the cladding and support structures. Careful "aerodynamic"
studies and design changes have minimized this effect in this class of reactor.

TABLE 9.1. Fuel-element Failures in CEGB Stations

Type	Number
Weld and end cap leaks due to manufacturing defects and handling damage	68
Body failures	1054
Irradiation growth	21
The loading of HT elements into LT positions and *vice versa* by accident or design	26
Identified in-core handling damage	34
Unknown causes	31

Source: J. Gittus, *Physical Metallurgy of Reactor Fuel Elements*, The Metals Society, 1975.

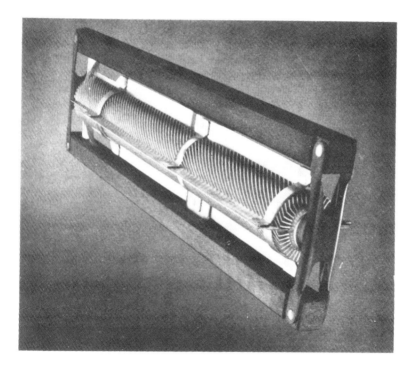

Fig. 9.3. Improved design of Berkeley fuel element.
Courtesy: UKAEA. Source: M. Simnad.

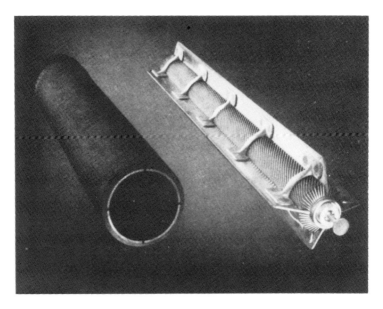

Fig. 9.4.   View of a Hunterston fuel element with sleeve.
Courtesy: UKAEA. Source: M. Simnad.

FRENCH GAS-GRAPHITE

French experience with Magnox-type reactor fuel elements has paralleled the
UK experience fairly closely although the materials have differed slightly [4].
France built three production reactors G1, G2 and G3 at Marcoule followed by
seven power reactors at Chinon 1, 2, 3, Saint Laurent 1 and 2, Bugey 1 and
Vandellos, the last being a joint venture with Spain. Subsequent French
reactors were PWRs. The characteristics of the seven power reactors are
given in Table 9.2. Design changes occurred from reactor to reactor.
Chinon 1 used a natural uranium-0.5% Mo alloy in the form of cast hollow rods
3.5 cm o.d. × 1.4 cm i.d. These were 50 cm long and were clad in a Mg-0.6% Zr
can, 1.8 mm wall thickness, with 24 longitudinal fins. The fuel rod diameter
increased up to 4.3 cm o.d. × 2.5 cm i.d. in Chinon 3 and Saint Laurent 1 and
after Chinon 1 the fuel composition was changed to U-1.1% Mo to prevent creep
collapse of the hollow rods.

In SL-1, SL-2 and Vandellos the fuel rod has a graphite core formed by casting the metal around the graphite as shown schematically in Fig. 9.5.
Bugey 1, the last reactor to be commissioned (in 1972) has an annular fuel
element design, i.e. the coolant flows over the inside as well as the outside
of the fuel rod giving a higher specific power (Fig. 9.6). The fuel is the
U Sicral F1 alloy, containing 0.07 wt % Al, 0.03 wt % Fe, 0.01 wt % Si and
0.01 wt % C, similar in composition and properties to the UK adjusted uranium.
The Mg-0.5% Zr cladding is metallurgically bonded to the fuel by a layer of
aluminium to overcome developmental problems due to nonbonding. The aluminium
diffuses into the cladding during irradiation and sets a limit on fuel
performance, determined by the reaction layer thickness which in turn is
determined by time and temperature. After problems with plutonium diffusion
through the cladding in Chinon 1, graphite or aluminium diffusion barriers
were incorporated in all fuel elements.

TABLE 9.2. French Gas-cooled Reactors

Reactor		CH 1	CH 2	CH 3	SL 1	SL 2	Vandellos	Bugey 1
Date commissioned		06/63	03/65	10/67	03/69	08/71	05/72	04/72
Net power	MW	70	210	460	480	515	480	540
Power generated on Aug. 31, 1974	GWh	2510	10,850	8405	10,310	9130	6523	5270
Thermal power	MW	240	800	1560	1660	1800	1750	1950
Total number of days at full power	JEPP	1740	2400	860	973	816	556	475
Loaded channels		1130	1980	2737	2934	2904	2916	852
Number of fuel elements		16,700	23,700	41,055	44,000	43,200	43,200	12,800
Average specific power	MW/MT	3.0	3.2	3.4	3.4	3.7	3.7	7.0
CO_2 pressure	bar	25	26.5	26.5	26.5	28.5	28.5	43.0
Max cladding temperature	°C	490	500	500	500	510	510	510
Max uranium temperature	°C	620	620	620	620	640	640	640
CO_2 flow per channel	kg/sec	1.5	2.4	3.15	3.15	3.6	3.6	13
Max burn-up	MWd/MT	4000	5000	5000	5000	6500	6500	5000
Average burn-up of unloaded fuel	MWd/MT	2700	3500	3500	3500	4700	4700	3500
Average burn-up with axial permutation	MWd/MT		4400	4400	4400	5900	5900	
Fuel consumption:								
tubular elements		47,500	67,000	83,500	58,500	64,500	58,000	
graphite core element			23,000	8500	24,000			
annular element								21,600
Failed elements								
nominal fuel		12	2	10	8	–	–	–
experimental fuel		3	3	–	–	–	–	–

Source: B. Boudouresque, *Trans. ANS* **20**, 279 (1975).

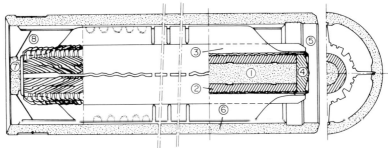

Cartouche

(1) Noyau graphite – graphite nut
(2) Tube uranium ∅ 23 x 43 – 10,250 kg
(3) Gaine magnesium (0.4% Zr). Magnesium can
(4) Bouchon magnesium (1.2% Mn). Magnesium cap
(5) Soudure tig – tig welding

Chemise

(6) Chemise graphite (∅ ext. 137 – L. 600 mm) – graphite sleeve
(7) Sellette support – supporting saddle
(8) Fil inox (∅1.5 mm) – stainless steel wire

Fig. 9.5.   French gas-cooled reactor fuel element.
<u>Source</u>: Trans. ANS, Vol. 20 (1976).

As with the Magnox fuel elements, the French fuel elements are held in graphite sleeves which facilitate the loading and unloading process (see Fig. 9.5). The short assemblies are loaded and unloaded while the reactors are at power. There are generally 15 assemblies (elements) stacked in each coolant channel. The CO_2 coolant enters at about $200^\circ C$ and exits at $400^\circ C$. It will be seen in Table 9.2 that the maximum fuel temperature is $640^\circ C$ — very close to the α/β transformation temperature, which was thought at one time to be bad practice. However, the failure rate of elements to peak burn-ups close to 7000 MWd/t(U) has been very low. Most failures were due to damage caused during fuel handling. Failures external to the fuel rod were caused by the fatigue fracture of lugs that position the rod in the graphite sleeve. The source of the fatigue stresses was the disturbance of the cool-ant flow; redesign of the lugs removed this problem.

Provided that a number of fairly simple rules are obeyed (with respect to the fuel fabrication procedure, fuel-clad bonding, clad sealing, thermal cycling and maximum fuel temperature) the metal fuel rods in the UK and French gas-cooled reactors behave extremely well. With their on-load fuel changing, these reactors have consistently achieved plant factors in excess of 90%; a great virtue during severe winters, coal and rail strikes and other natural hazards. This feature compensates for the relatively low thermal efficiency and low burn-up of the fuel.

AGR

The AGR was conceived as a logical step in the development of gas-cooled thermal reactors [5] The temperature limitations of uranium metal were removed by changing to UO_2, which had proved to be successful in the early water reactors in the USA and Canada. CO_2 was retained as the coolant, but the magnesium cladding was abandoned because of its low melting point ($\sim 650^\circ C$). For a short time beryllium was considered for the cladding, but attention turned to an Fe-20% Cr-25% Ni-0.5% Nb alloy that had good strength and corrosion resistance at high temperatures.

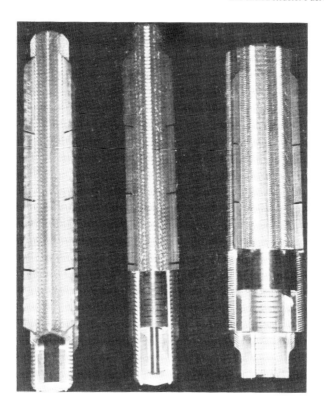

Fig. 9.6.   Types of fuel element.   Source: Physical
Metallurgy of Reactor Fuel Elements.

Much of the AGR fuel element testing has been performed in the Windscale AGR
(WAGR), a 100 MWt, 32 MWe power prototype.   As of 12/31/79 two commercial
AGR plants were in operation in England and two in Scotland, with several
others at advanced stages of construction.   The AGR retained the concept
used in the Magnox reactors of a stringer of several (usually 4) fuel element
assemblies per channel.   Hence the fuel rods are fairly short and have little
or no room for a fission gas plenum.   Consequently the fuel has to be operated
in a regime of low fission gas release.   The fuel is of low enrichment (2-3%)
and is designed for a peak burn-up approaching 50,000 MWd/t(U).

The "driver" fuel for WAGR was about 0.51 m (20 in) long with 0.38 mm thick
cladding (15 mil) over UO_2 pellets 10.2 mm (0.4 in) diameter.   The cladding
was ribbed to promote better heat transfer, the rib size being ~6 mil square
on a 0.1 in pitch.   Twenty-one fuel tubes were assembled into a cluster,
held by stainless steel grids within a graphite sleeve.   A central steel tie-
rod held adjacent clusters together (Fig. 9.7).

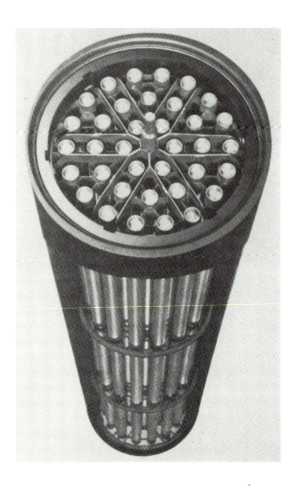

Fig. 9.7.   Cutaway of a 36 pin AGR fuel assembly. This
            type of element is used in the Dungeness B
            reactors. Courtesy: UKAEA. Source: M. Simnad.

These Mk II elements performed well with a clad surface temperature of 650°C
to burn-ups as high as 31,000 MWd/t(U). However, the design was not economic.
Furthermore, there was an increasing need for the UK nuclear stations to be
capable of operating in a load-following mode, as opposed to their traditional
role of base-load stations. Accordingly, the Mk III and Mk IV fuel elements
were designed with larger diameter rods (pellet diameter = 14.5 mm (0.58 in)).
About 10,000 pins of this design were irradiated in WAGR with a maximum
cladding temperature of 850°C (designed to give maximum steam cycle efficiency
of ≥40%). When subjected to a sudden increase in power rating the pins tended
to fail. The failure susceptibility is plotted in Fig. 9.8; $Y(\varepsilon)$ is a rating
parameter that is proportional to the magnitude of the ramp increase in
rating plus the value of the subsequent mean rating. The failures were
characterized by a crack in the cladding opposite radial cracks in the fuel
(Fig. 9.9).

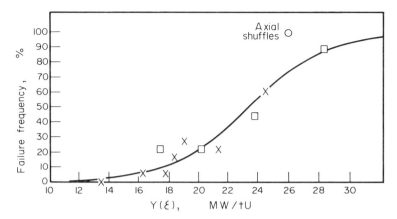

Fig. 9.8. Failure frequency of Mk III and Mk IV fuel pins
v. a rating parameter. Source: J. Gittus,
Phys. Met. Reactor Fuel Elements.

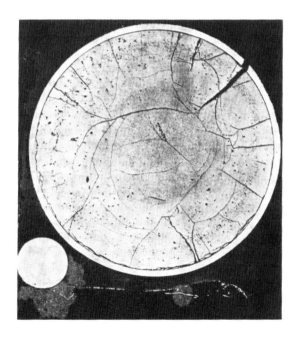

Fig. 9.9. 2 Coincident cracks in pellet and clad.
Source: J. Gittus.

Briefly, the mechanism of failure was discovered to be due to a combination of circumstances. Prior to the ramp, the fuel pellet was cracked (due to thermal stresses) and the cladding was in intimate contact with the fuel because the thin cladding creeps under the external pressure of the CO_2 coolant. The fuel was designed to operate at a temperature that was low enough to retain the fission gases in solution or as very small bubbles. When the ramp occurred, the fuel centre temperature rose and the fission gases rapidly migrated to the grain boundaries of the UO_2 to form bubbles and to cause the fuel to swell. The swelling stresses were transmitted to the cladding and were most severe at the ends of radial cracks. The cladding had low ductility due to its fabrication history and to radiation-embrittle-ment (helium bubble formation on grain boundaries). Furthermore, fission products may have contributed to crack initiation at the ends of the fuel cracks. In any event, the sudden application of an increased, localized stress led to failure.

The solution was fairly simple: to substitute hollow pellets for the solid pellets. This provides expansion room in the centre of the fuel to accommodate the ramp-induced swelling. A further improvement was made by modifying the cladding. Instead of using cold-worked cladding (which has a low creep-rupture ductility) an additional heat treatment was introduced to cause recrystallization or over-aging, resulting in a substantial increase in tube-burst ductility. A large number of hollow fuel rods with this modified cladding have been irradiated in WAGR with a very low failure rate. Some clusters have been irradiated in loop experiments where the power was cycled several times a day. While 50-150 cycles caused the original fuel to fail, the modified fuel has survived 740 cycles in which the power was changed by 25%.

This form of fuel has been adopted as the standard for commercial AGRs, the fuel rod length having been increased to 1.0 m (39 in).

HTGR

The high temperature gas-cooled reactor is the most advanced form of commercial gas-cooled reactor. It differs from the other gas-cooled systems in that it uses helium as the coolant and the fissile and fertile materials are dispersed throughout the graphite moderator.

The dispersion of the fuel and fertile material in the form of particles $\sim$500 μm (20 mil) in diameter has many advantages in terms of lower mean power density in the core, less concern with failed fuel rods, and the possibility of high temperature operation. The latter includes the possibilities of direct-cycle gas turbine operation and process heat supply.

While there are many novel features in the HTGR system, the most important is the coating of each fuel or fertile material particle to isolate it from the moderator and coolant [6,7]. The fuel may be UC_2 or UO_2; General Atomics Company (GAC) favour the former, while UO_2 is favoured in Europe. The fuel particles are fabricated by standard processes, which may include the sol-gel route. The particles are levitated in a fluidized bed by an upward flowing inert gas. Hydrocarbons are introduced at certain temperatures to form carbon layers on the fuel by pyrolysis (Fig. 9.10). Varying the temperature can produce an initial low density (buffer) layer which accommodates swelling and fission product recoils, followed by denser sealing layers that perform the role of cladding, i.e. a mechanical, gas-tight seal. Two graphite layers constitute a BISO particle, which is generally adequate for use in typical

large, electricity-producing HTGRs with fuel temperatures generally below
1000°C. At higher temperatures fission products, especially caesium, migrate
through the graphite layers into the graphite and hence into the coolant.
This is solved by the TRISO particle in which an intermediate layer of SiC
is deposited by introducing silane into the fluidized bed coating furnace.
This provides protection up to ∿1500°C.

Mention has been made of fertile particles. The HTGR is particularly suited
to the ^{232}Th/^{233}U cycle, although it can operate on several others. Normal
practice to date has been to start reactors with highly enriched (in ^{235}U)
uranium and to irradiate ThO$_2$ particles to generate ^{233}U. These are also
coated (with carbon) and are usually dispersed in different graphite rods to
make reactor control and reprocessing easier. In passing, it should be noted
that a low enrichment version has been developed that can be a once-through
system, i.e. consistent with nonproliferation policies (Table 2.5).

The fissile and fertile coated particles are dispersed in a graphite matrix
by various methods, e.g. GAC uses an injection moulding process in which beds
of particles are bonded with a viscous mixture of pitch and graphite powder
and then heated. The objective is to form the particles and graphite into
a solid or hollow rod of graphite or into balls for the German pebble-bed
design.

In the 340 MWe Fort St. Vrain reactor built by GAC for the Public Service
Company of Colorado, the "fuel element" consists of a hexagonal block of
machined, high-density graphite 36 cm (14 in) across the flats x 80 cm
(31 in) long. Three hundred and eighteen vertical holes ∿1.27 cm (0.5 in)
diameter are drilled to accommodate coolant in some (108) and fuel rods in
others (210) (Fig. 9.11). These hexagonal blocks are stacked eight blocks
high in 493 vertical columns to form an active core 19.5 ft diameter x 15.5 ft
high (Fig. 9.12).

The integrity or performance of the HTGR fuel elements depends on the
behaviour of the coated fuel particles and on the behaviour of the graphite
blocks, which are to some degree independent of one another.

The factors that determine the life to failure of coated particles are
mechanical and thermochemical. Computer codes have been developed to model
the coating stresses as a function of fuel lifetime or burn-up. While early
emphasis was on the behaviour of the graphite layers in the BISO particle,
current emphasis is on the properties of the SiC layer, since this is
stronger at the operating temperature. A given batch of coated particles
will exhibit a spread of particle diameters and buffer layer thicknesses.
These two factors determine the fission gas pressure within the particle.
The other layer thicknesses are reasonably constant. Hence one can construct
a plot of the range of particle diameters and buffer layer thicknesses and
compare these with the SiC layer end-of-life stresses (Fig. 9.13). Experience
has shown ∿30,000 psi to be the failure threshold for SiC. Hence, the batch
depicted in Fig. 9.13 should exhibit a low failure rate, as indeed seems to
be the case in practice. One has to remember that a large HTGR core contains
∿10^{11} particles, hence even at a low failure rate (say 0.1%) a significant
number (10^8) will fail. What will be important will be the geometrical and
time distribution of these failures.

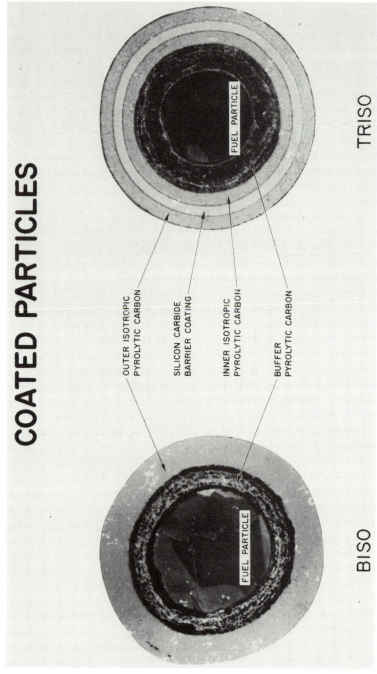

Fig. 9.10. Coated particles. Courtesy: General Atomic Corporation.

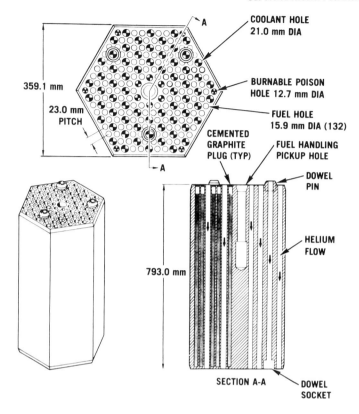

COOLANT HOLE
21.0 mm DIA

BURNABLE POISON
HOLE 12.7 mm DIA

FUEL HOLE
15.9 mm DIA (132)

CEMENTED
GRAPHITE
PLUG (TYP)

FUEL HANDLING
PICKUP HOLE

DOWEL
PIN

HELIUM
FLOW

359.1 mm

23.0 mm
PITCH

793.0 mm

SECTION A-A

DOWEL
SOCKET

Fig. 9.11. Assembly of an HTGR fuel element. Courtesy:
General Atomic Corporation.

The thermochemical process that limits coated particle life is termed the
amoeba effect, because the optical microscope sections are reminiscent of an
amoeba moving through water. The amoeba effect is the movement of the fuel
particle up a temperature gradient relative to the coatings and matrix
(Fig. 9.14). It has been observed with carbide and oxide fuels, but the
mechanisms are different.

The amoeba effect is basically the thermal migration of carbon down the
temperature gradient, i.e. the buffer layer "dissolves" on the hot side of
the particle (or kernel), carbon is transported down the gradient and rejected
on the cooler side. This can lead to particle failure through the "corrosion"
of the load-bearing layers. The process in carbide fuels (UC_2) and fertile
particles (ThC_2) has been shown to be controlled by the solid state diffusion
of carbon in the compound. It is unaffected by radiation so that out-of-pile
thermal gradient diffusion tests can be used to define the conditions for
avoiding the effect.

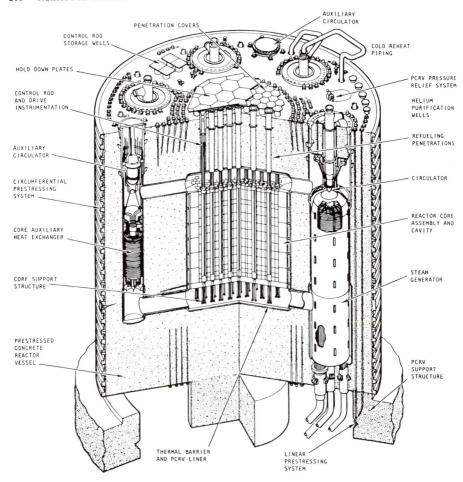

Fig. 9.12.   HTGR prestressed concrete reactor vessel
arrangement.   Courtesy: General Atomic Company.

The primary system components are contained in a large
cylinder of prestressed concrete. Penetrations exist for
refueling, as well as for servicing (and even replacing)
various pieces of equipment. Several primary coolant
loops, as well as secondary cooling loops, are contained
in the vessel.

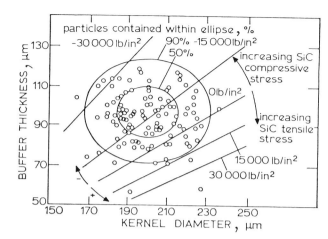

Fig. 9.13. Two-dimensional distribution of kernel diameter and buffer thickness for a TRISO-coated particle batch with lines of constant end-of-life SiC stress. Source: Gulden, Phys. Met. of Reactor Fuel Elements.

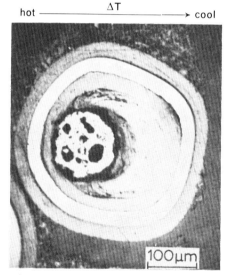

Fig. 9.14. An extreme example of amoeba migration in irradiated UO_2. Source: Gulden, Phys. Met. of Reactor Fuel Elements.

In oxide fuel (UO_2) the most likely carbon transport mechanisms are either CO/CO_2 reactions (i.e. gas phase transport) or oxygen ion diffusion through the UO_2. As UO_2 fissions, the oxygen partial pressure within the particle will change. ^{235}U fission results in the generation of excess "unbound" oxygen that is available to transport carbon via the reaction $C + CO_2 \rightleftharpoons 2CO$. In ThO_2 the rate is much lower than in UO_2, possibly because ^{233}U fission later in life in this type of particle tends to liberate less oxygen than ^{235}U fission due to the different fission product yield spectrum.

While the cause of the amoeba effect in UO_2 particles is not fully understood, the cure *is* understood. The incorporation of an oxygen getter in the particle, to prevent the formation of CO during irradiation, should greatly reduce the effect.

Finally, the integrity of graphite bodies in a reactor core is related to their dimensional stability under irradiation. At high temperatures, graphite tends to shrink under irradiation. Since graphite crystals are anisotropic, their shrinkage is anisotropic. Consequently graphite blocks must be made to have a random orientation of the crytals. As in metals, radiation growth or shrinkage is offset by radiation creep. Solutions to these problems in power reactor graphite bodies have been largely empirical, supported by a range of basic radiation effects data. In general, graphite is a robust "forgiving" material and the incidence of failures in graphite bodies in HTGRs has been very low. In any case, the cracking of a graphite block is not serious unless it impedes fuel changing or it causes particle failures, neither of which is very probable.

SAFETY

We dealt at some length in the previous chapter with reactor safety questions in relation to fuel elements. The advanced gas-cooled reactors (AGRs and HTGRs) have very different safety considerations from the water reactors. First, the graphite moderator is a large heat sink. Second, the materials of the fuel elements are different: graphite has a very high melting point and stainless steel has good high temperature corrosion resistance. Most significant, however, is the use of prestressed concrete reactor vessels (PCRVs) which are massive and from which a large leak is highly improbable. Auxiliary cooling loops, using diesel power if necessary, are adequate to remove fission product decay heat.

REFERENCES

1. Jepson, M. D. and Slattery, G. F., British Patent No. 863, 492, 1958/61.
2. Greenfield, P., *et al.*, A/CONF 28/P146 (May 1964) and Stewart, J. C. C., A/CONF 28/P560 (May 1964).
3. Eldred, V. W., Harris, J. E., Heal, T. J., Hines, G. F. and Studdard, A., in *Physical Metallurgy of Reactor Fuel Elements*, The Metals Society, pp. 341-351 (1975).
4. Blanchard, P., Millet, P., Courcon, P., Truffert, J. and Faussat, A., ibid, pp. 352-364.
5. Gittus, J. H., ibid, pp. 369-373.
6. Price, M. S. T. and Shepherd, L. R., ibid, pp. 397-409.
7. Gulden, T. D., Harmon, D. P. and Stansfield, O. M., ibid, pp. 410-415.

CHAPTER 10

Fast Reactor Fuel Elements

INTRODUCTION

Fast breeder reactors have been under development since the mid 1940s — earlier than water- and gas-cooled thermal reactor development. The world's first 1000 MWe fast breeder reactor, Superphenix, is due to operate in 1983 and the first truly commercial stations should be operating in the early 1990s. This is a long development period, caused by the considerable technical difficulties that have had to be overcome. Not the least of these is the design and performance of the fuel elements.

A fast breeder reactor is an unmoderated reactor with a fairly highly enriched core. The physics and fuel cycle economics of the system dictates that it is compact, to avoid neutron leakage, and it operates with a high power density to high fuel burn-ups. The universal choice of coolant has been liquid sodium because it has good heat transfer properties and is compatible with cladding and structural materials.

A key part of the fast breeder reactor concept is breeding. ^{239}Pu produces on the order of 2.9 neutrons per fission of ^{239}Pu in a fast spectrum. Thus 1.9 neutrons are "spare" and may be absorbed in ^{238}U to produce more plutonium. In practice, of course, some neutrons are absorbed in structural materials, but significantly more than 1.0 per fission event are available to breed more plutonium. The breeding ratio is the ratio of bred plutonium to the core investment of plutonium, in steady state. Thus, a fast breeder reactor that has been operating long enough to be in equilibrium may produce $\sim$1.6 times the plutonium burned in the core, i.e. sufficient for a core reload plus 0.6 of core spare with which to fuel new reactors. The term "doubling time" is also used to denote the time in years that it would take for a fast breeder to generate two core loadings. Breeding ratios for a fast reactor with different fuels are listed in Table 10.1

There are two possible breeding cycles: ^{238}U $\rightarrow$ ^{239}Pu and ^{232}Th $\rightarrow$ ^{233}U. However, only the former is seriously considered for fast reactors. Thus a breeder reactor core is surrounded by a blanket of breeder elements which convert ^{238}U to ^{239}Pu by neutron capture.

TABLE 10.1.   Comparative Characteristics of Various Fuels in a Reactor with Core Volume 1000 l, Thermal Capacity 800 MW, and Core Composition 30% Fuel, 20% Stainless Steel, and 50% Sodium (H/D = 1)

Parameters	Units	Fuel				
		U–Pu–Mo	PuC–UC	PuO_2–UO_2	PuO_2– stainless steel	PuO_2–U + 10 wt% Mo
1. Composition		10.3 wt% Pu	14.06 vol.% PuC	16.39 vol.% PuO_2	12.75 vol.% PuC_2	19 vol.% PuO_2
2. Critical mass	kg ^{239}Pu	519	492	447	548	518
3. Breeding ratio		1.77	1.66	1.57	1.32	1.68
4. Core power	MW	716	712	702	671	723
5. Average power density	kW/kg ^{239}Pu	1541	1626	1790	2299	1544
6. Number of fuel elements in core[a]		23,964	3267	20,968	21,680	14,738

[a]The diameter of the fuel elements was calculated on the basis of maximum temperature at the centre of the element.

Source: M. Simnad.

Fast reactor fuel elements have a fissile content that varies with reactor size. Thus small reactors like EBR-II and DFR have a fully enriched fuel, while a 1000 MWe, 3000 MWt reactor is likely to have an enrichment of less than 20%. Nevertheless, this is much higher than in a thermal reactor. This large fissile investment combined with the high power density requires the fuel to operate to much higher burn-up levels than in thermal reactors. Furthermore, the breeding process counteracts loss of reactivity with burn-up. A typical burn-up target for fast reactors is 10% of heavy atoms or $\sim$100,000 MWD/te. This gives reasonably long times between refuelling. The high power density, however, produces a high fast neutron flux which generates radiation damage at a high rate, and it is this factor that may restrict the life of the fuel elements due to the degradation of the properties of the cladding and duct materials.

Fast reactor fuel elements are smaller than thermal reactor elements (Fig. 10.1). The smaller core dimensions give a shorter fuel column length (typically 3 ft or 1 m) and the high power density gives a smaller fuel diameter (in order to keep fuel centre temperatures well below their melting points). The characteristic features of a fast reactor fuel pin are shown in Fig. 10.1. At each end of the fuel column is an axial blanket of uranium and the fuel pin has a fairly large plenum to accommodate the large quantity of fission gases that are released in going to high burn-up at a high fuel centre temperature.

Because fast reactor fuel pins are small in diameter, it is generally convenient to group them into subassemblies which are hexagonal in cross-section (e.g. see Fig. 10.3). These stack together to form a reasonably cylindrical and compact core. The pins may be spaced — to form regular coolant channels — by wires wrapped around them or by grids. All of these components receive a high fast neutron dose during their life in the core. In a 1000 MWe reactor the peak flux will be around 10^{16} n/cm^2/sec and the peak fluence or dose around 10^{23} n/cm^2/sec. We have discussed the changes in properties that this can cause in Chapter 4.

In a breeder reactor the total fuel cycle is very important. The core and blanket elements must be reprocessed to separate out the fission products, the uranium and the plutonium, the latter being added to the new fuel elements that are built. At equilibrium the recycled plutonium contains active isotopes producing neutrons and gammas. Therefore, refabrication has to be performed remotely behind shielding. This aspect has yet to be fully demonstrated.

EARLY REACTORS (Table 10.2)

The first generation of fast breeder reactors were small in size — between 1 and 60 MWt — and were built to test the physics and engineering of this class of reactor. The first of these was the Experimental Breeder Reactor-1 or EBR-I built by Argonne National Laboratory at its Idaho site, with initial operation in 1951. In December 1951 the reactor was coupled to a turbo-alternator and produced the world's first electricity from nuclear power, only nine years after Fermi's first demonstration of the chain reaction.

TABLE 10.2. Major Fast Reactors

Reactor name	Location	Type	Output	Fuel	Coolant	Start-up/ shutdown dates
USA						
EBR-1 (Experimental Breeder Reactor-1)	NRTS, Idaho	First generation fast breeder	1.4 MW(t) 150 kW(e)	Plutonium	Sodium-potassium	1951 start-up 1964 shutdown
EBR-2 (Experimental Breeder Reactor-2)	NRTS, Idaho	First generation fast breeder	62.5 MW(t) 16.5 MW(e)	PuO_2/UO_2	Sodium	Nov. 1963 start-up
EFFBR (Enrico Fermi Fast Breeder Reactor)	Lagoona Beach, Mich.	First generation fast breeder	200 MW(t) 66 MW(e)	PuO_2/UO_2	Sodium	1963/1966
FFTF (Fast Flux Test Facility)	Richland, Washington	Second generation experimental fast reactor	400 MW(t) 0 MW(e)	PuO_2/UO_2	Sodium	1982
SEFOR (Southwest Experimental Fast Oxide Reactor)	Stickler, Arkansas	Second generation experimental fast reactor	20 MW(t) 0 MW(e)	PuO_2/UO_2	Sodium	1969
FARET (Argonne Fast Reactor Experiment Test)	NRTS, Idaho	Experimental reactor to test fast power reactor components	50 MW(t)	Various	Sodium	1965 (project termination)
CRBR (Clinch River Breeder Reactor)	Oak Ridge, TN	Second generation fast breeder prototype	1000 MW(t) 350 MW(e)	PuO_2/UO_2	Sodium	Start-up ~1989

Reactor name	Location	Type	Output	Fuel	Coolant	Start-up/ shutdown dates
USSR						
BR-1	Obnisk	First generation fast breeder	0 MW(t) 0 MW(e)	Plutonium metal	–	1955 start-up 1956 shutdown
BR-5	Obnisk	First generation fast breeder (based on BR-1)	5 MW(t) 0 MW(e)	PuO_2	Sodium	June 1958 start-up
BOR-60	Melekess Ulyanovsk Region	Second generation experimental fast reactor	60 MW(t) 12 MW(e)	PuO_2/UO_2 or UO_2	Sodium	Dec. 1968 start-up
BN-350	Shevchenko (Caspian Sea)	Second generation fast breeder reactor	1000 MW(t) 350 MW(e)	PuO_2/UO_2 or UO_2	Sodium	1969 start-up
BN-600 (Beloyarsk-III)	Sverdlovsk Region	Fast breeder	600 MW(e)	–	Sodium	1980
United Kingdom						
DFR Dounreay Fast Reactor	Dounreay	First generation fast breeder	72 MW(t) 14 MW(e)	Uranium metal (45.5% enriched)	Sodium-potassium	Nov. 1959 start-up 1979 shutdown
PFR (Prototype Fast Reactor)	Dounreay, Scotland	Second generation fast breeder prototype	600 MW(t) 250 MW(e)	PuO_2/UO_2	Sodium	1974 start-up
Germany, Federal Republic						
KNK	Karlsruhe	Second generation experimental fast reactor	58 MW(t) 20 MW(e)	–	Sodium	1971 start-up
SNR (Schneller Natrium-gekuehlter Reaktor)	Kalkar/Rhein	Second generation fast breeder	730 MW(t) 300 MW(e)	PuO_2/UO_2	Sodium	1980's

Reactor names	Location	Type	Output	Fuel	Coolant	Start-up/shutdown dates
Japan						
JOYO (Japanese Experimental Fast Reactor)		Second generation experimental fast reactor	92 MW(t) 0 MW(e)	PuO_2/UO_2	Sodium	1980 start-up
France						
RAPSODIE (Cadarache Reactors)	St. Paul-les-Durance	First generation fast breeder	20 MW(t) 0 MW(e)	PuO_2/UO_2	Sodium	Jan. 1967 start-up
PHENIX	Marcoule	Second generation fast breeder prototype	600 MW(t) 250 MW(e)	PuO_2/UO_2	Sodium	1973
Superphenix	Creys Murville	Fast breeder power reactor	3000 MW(t) 1200 MW(e)	PuO_2/UO_2	Sodium	1982
Italy						
PEC	Brasinone near Bologna	Second generation experimental fast reactor	130 MW(t) 0 MW(e)	UO_2	Sodium	?

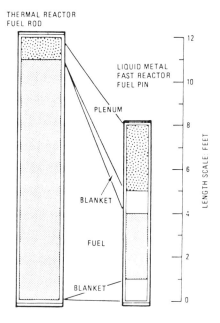

DESIGN PARAMETERS

LENGTH (ft)	∿12	6 TO 8
DIAMETER (in.)	0.5	0.25 TO 0.30
FUEL COLUMN LENGTH (ft)	∿11	3 TO 4
BLANKET LENGTH (in.)	NONE	∿10 TOP AND BOTTOM
RATIO OF PLENUM TO FUEL COLUMN	0.08	1 TO 1.5
CLADDING MATERIAL	ZIRCALOY	STAINLESS STEEL

Fig. 10.1. Thermal and fast reactor fuel design
parameters. Source: M. Simnad and
J. Howe, Mat. Sci. in Energy Tech.
CH2, 1979.

EBR-1 had a very small core: 8.5 in high and about 10 in across. It generated
1.4 MWt, 0.2 MWe with NaK coolant entering the core at 230°C and exiting at
322°C. The initial fuel elements were single rods (Mk I and Mk II) of fully
enriched cast uranium 0.364 in (0.92 cm) (Mk I) and 0.384 in (0.95 cm) (Mk II)
in diameter. The vacuum induction melted and cast bars were hot rolled at
300°C to rods with beta anneals at 725°C to refine the structure. For the
Mk II core a U-2% Zr alloy was centrifugally cast into rods. The rods were
NaK bonded in Type 347 stainless steel tubes which were loaded individually
in the core.

During operation of the Mk II core with reduced coolant flow the fuel
elements bowed towards the core centre, giving a positive reactivity
coefficient that led to partial melting of the core. A heroic effort was
mounted to dismantle and analyse the core [1]. After this the core was
redesigned to group the fuel elements into subassemblies that provided more

rigidity and resistance to bowing. The Mk III element (Fig. 10.2) contained
U-2% Zr fuel (as in Mk II), but inside a Zircaloy-2 tube to which it was
metallurgically bonded. Blanket rods were made of natural uranium-2% Zr rods.
The subassemblies (Fig. 10.3) were 2.875 in (0.72 cm) across the flats and
contained 36 rods with a tightening rod in the centre. The layout of the
Mk III core is shown in Fig. 10.4.

In 1962 a plutonium core was loaded containing Pu-1.25 wt % Al or Pu-10 at %
Al cast rods NaK bonded inside Zircaloy-2 cladding. The fuel was found
to creep significantly above 550°C, so that operation of the fuel was
restricted to a maximum temperature of 450°C. The fuel only operated for
less than a year to a burn-up of only 0.07%; however, very useful physics
data was generated.

The Russian BR-5 reactor went critical in 1959 [2]. It was a 5 MWt sodium-
cooled reactor fuelled initially with PuO_2 pellets in Type 321 stainless
steel cladding, with helium bonding. In 1965 UC pellet fuel was used. The
fuel pins were small — 0.197 in (0.5 cm) o.d. and 0.16 in (0.4 cm) fuel
diameter. The rods were placed in groups of 19 inside subassemblies 1.023 in
(2.6 cm) across the flats. The PuO_2 fuel was taken to 5% burn-up, but
numerous failures occurred resulting in primary circuit contamination [3].
In fact, a lot was learned about operation with failed fuel elements. The
UC and UO_2 fuel elements also failed — mainly due to fuel swelling stresses
on embrittled cladding.

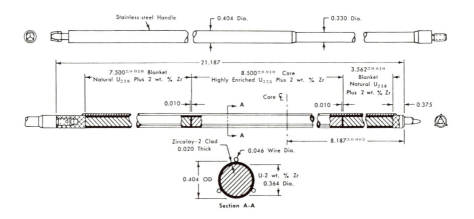

Fig. 10.2. Fuel rod for EBR-I Mk III. Dimensions are
in inches. Source: A. R. Kaufmann, Nuclear
Reactor Fuel Elements, Interscience 1960.

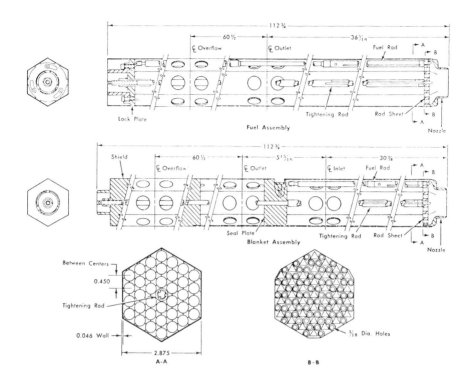

Fig. 10.3.   Core and inner-blanket subassemblies for
             EBR-I Mk III.   Dimensions are in inches.
             Source: A. R. Kaufmann.

The British DFR (Dounreay Fast Reactor) was a 60 MWt, 12 MWe NaK-cooled fast
reactor which went critical at Dounreay, Scotland in November 1959.   The fuel
elements were designed to be loaded individually and not in a subassembly.
The fuel element was designed to give protection against a loss of flow type
of accident.   It was a hollow rod of U-2$\frac{1}{2}$ a/o Cr (Mk I and Mk II), U-20 a/o
Mo (Mk III) and U-7 a/o Mo (Mk IV).   The idea of the hollow rod was to clad
the central hole with a metal that reacted with the fuel at a lower
temperature than was the case at the outer diameter.   Experiments on uranium-
metal reactions led to the choice of vanadium for the inner cladding and
niobium for the outer [4].   The reason is shown in Fig. 10.5, which shows the
can penetration temperatures for the two metals.

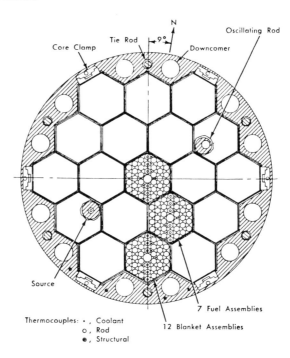

Fig. 10.4.   Cross section through centre line of EBR-1
Mk III core.   Source: A. R. Kaufmann.

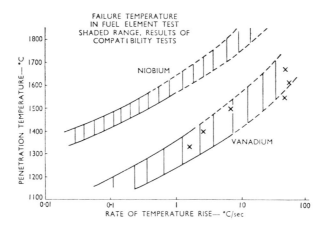

Fig. 10.5.   Can penetration temperatures based on average
attack.   X failure temperature in fuel
element test.   Shaded range covers results of
compatibility tests.   Source: C. R. Tottle
et al., UKAEA.

The design of the DFR element is shown in Fig. 10.6 [5]. The element was 48.35 in in overall length. The fuel length was 21 in (53 cm). It was open to the NaK coolant via vent holes and was sodium-filled during fabrication. The outer 0.020 in (0.05 cm) thick niobium cladding had spiral fins to maintain coolant flow if bowing occurred. The bottom end of the element was fashioned into a spike to permit easy location in the bottom grid. The core was assembled, as shown in Fig. 10.7, to a star pattern that was surrounded by natural uranium blanket elements. The core diameter was 19.6 in ($\sim$50 cm).

In general, the behaviour of the U-7 a/o Mo fuel was satisfactory. The burn-up limit was about 2% and the maximum fuel temperature 680°C. Sufficient confidence was generated in fuel performance that later cores dispensed with vanadium cladding and used niobium both on the inside and the outside of the fuel.

A problem was found with the blanket elements which had distorted due to cycling through the uranium α-β transition at 660°C — caused by coolant flow instabilities. These elements were replaced and reactor operation continued.

From about 1965 to 1979, when it shut down, DFR was a valuable test bed for more advanced breeder reactor fuels and cladding. The centre of the core (see Fig. 10.7) was modified, so that an experimental hexagonal section was formed in which groups of oxide and carbide fuel elements (experimental subassemblies) were irradiated [6].

Chronologically, the next breeder reactor to consider is the Enrico Fermi Atomic Power Plant built at Monroe, Michigan, by Atomic Power Development Associates. It was an attempt by industry to speed commercialization of the breeder reactor by building a 200 MWt, 60 MWe sodium-cooled plant. The fuel was U-10 w/o Mo alloy, 30.5 in (78 cm) long and 0.148 in (0.378 cm) diameter at 25.6% enrichment. The fuel was metallurgically bonded by coextrusion to 5 mil thick zirconium cladding and mounted in 105 pin assemblies [7].

The reactor started up in August 1963 and operated successfully until October 1966 when a zirconium sheet, forming part of a molten core deflector, broke off and blocked the entrance to two subassemblies. These partially melted but did not cause progressive failure [8].

Finally, we come to EBR-II, the only one of this class of reactors still operating. It was conceived as a demonstration of the closed fuel cycle [9]. The reactor was located next to a Fuel Cycle Facility so that spent fuel elements were shipped into that facility for pyrometallurgical reprocessing and refabrication [10]. This, of course, influenced the choice of fuel composition. The pyrometallurgical process consists of melting the "hot" fuel in suitable crucibles. The volatile fission products are driven off and are captured and certain fission products react with the crucible material to form a slag. What is left is an alloy of uranium with "fissium" (Fs), a mixture of the more noble fission products, i.e. molybdenum, ruthenium, rhodium, palladium, zirconium and niobium. Over three core loadings of recycled fuel were run through the complete system. It was then shut down and U-5Fs elements are now made synthetically for driver fuel. A disappointment was the poor behaviour of U-Pu-Fs fuel, particularly with regard to compatibility with steel cladding. U-Pu-Zr metal fuels developed by ANL have shown better compatibility [11], but their suitability for the closed fuel cycle has not been demonstrated.

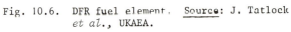

Fig. 10.6.   DFR fuel element.   Source: J. Tatlock
et al., UKAEA.

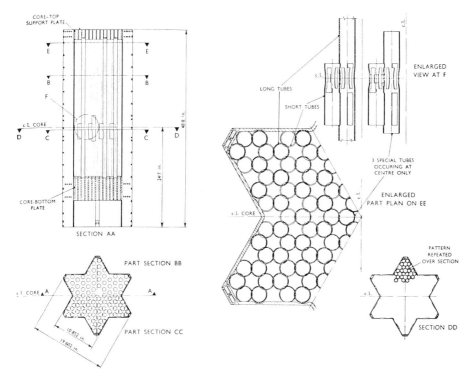

Fig. 10.7.   DFR core layout.   <u>Source</u>: J. Tatlock
et al., UKAEA.

Nevertheless, EBR-II has become the workhorse for US irradiations of fast
reactor fuels, cladding, structural materials and absorbers. It evolved,
like DFR, from a small demonstration plant to a test reactor. In doing so
the fuel element performance has improved greatly.

The EBR-II fuel pin is vacuum melted and injection cast into glass or silica
tubes to a precise dimension of 0.130 in (0.29 cm) diameter. 13.50 in (30 cm)
lengths of fuel are placed in 0.150 in (0.38 cm) i.d., 0.174 in (0.44 cm) o.d.
tubing of Type 316 stainless steel. The 0.010 in (0.025 cm) gap between fuel
and cladding is filled with a sodium bond which is carefully tested for
complete bonding. There is a 9.35 in (24 cm) gas plenum at the top of the
element (see Fig. 10.8). The 24 in (62 cm) long pins are loaded into a 91
pin subassembly and are spaced apart by spiral wound wire wraps. The subassemblies are 2.29 in (5.8 cm) across the flats and a core loading is made
up typically of 57 subassemblies along with perhaps 40 experimental fuel subassemblies and 25 nonfuel subassemblies. Figure 10.9 shows a typical core
loading plan.

The early Mk I and Mk Ia pins had a burn-up limit of about 2% set by the
swelling stresses on the cladding. However, it was realized that metal fuel
releases a lot of fission gas after it has swelled ∿25%. This is due to
linkage of the fission gas bubbles. Hence, the Mk II design allowed sufficient room for the fuel to swell more than 25% before it touched the cladding.
This Mk II design has operated to burn-ups as high as 16% without failure —
a great advantage in a test reactor.

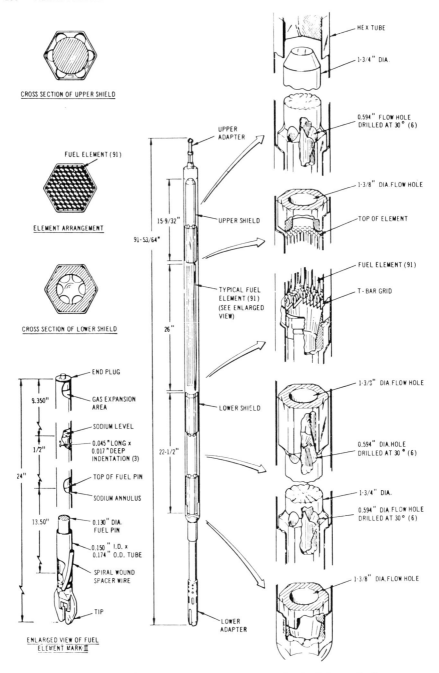

CROSS SECTION OF UPPER SHIELD

FUEL ELEMENT (91)

ELEMENT ARRANGEMENT

CROSS SECTION OF LOWER SHIELD

END PLUG

GAS EXPANSION AREA

SODIUM LEVEL

0.045" LONG x 0.017" DEEP INDENTATION (3)

TOP OF FUEL PIN

SODIUM ANNULUS

0.130" DIA. FUEL PIN

0.150" I.D. x 0.174" O.D. TUBE

SPIRAL WOUND SPACER WIRE

TIP

ENLARGED VIEW OF FUEL ELEMENT MARK-II

9.350"

1/2"

24"

13.50"

UPPER ADAPTER

UPPER SHIELD

TYPICAL FUEL ELEMENT (91) (SEE ENLARGED VIEW)

LOWER SHIELD

LOWER ADAPTER

15-9/32"

91-53/64"

26"

22-1/2"

HEX TUBE

1-3/4" DIA.

0.594" FLOW HOLE DRILLED AT 30° (6)

1-3/8" DIA. FLOW HOLE

TOP OF ELEMENT

FUEL ELEMENT (91)

T - BAR GRID

1-3/8" DIA FLOW HOLE

0.594" DIA. HOLE DRILLED AT 30° (6)

1-3/4" DIA.

0.594" DIA FLOW HOLE DRILLED AT 30° (6)

1-3/8" DIA. FLOW HOLE

Fig. 10.8.   Components of an EBR-II Mk II driver fuel subassembly.   Courtesy: Argonne National Laboratory.

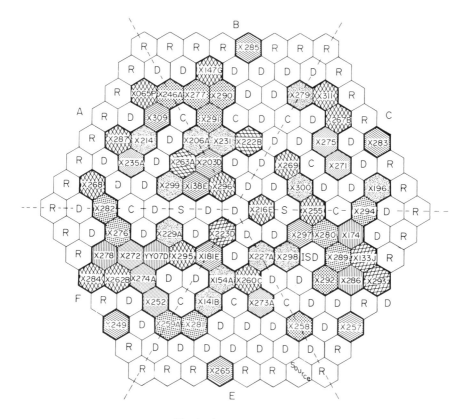

Start of run – June 19, 1977

Structural	D^*- Incot driver
Advanced fuels	D – Standard driver
Mixed oxide fuels	C – Control rod
Metal fuels	S – Safety rod
Absorber	C^*- Stainless steel rod
RBCB	ISD-Insat dummy
MISC.	R – Stainless steel reflector

Fig. 10.9.   EBR-II reactor – grid loading diagram.
Source: Argonne National Laboratory.

Metal fuels and blankets are attractive because they give the highest breeding
ratio (Table 10.1). They have the highest density of fissile atoms and contain no moderating atoms like oxygen in UO_2. However, they have relatively
low melting points and (worse) they react with stainless steel cladding to
form low melting point eutectics ($\sim 500^\circ C$). U-Pu-Zr alloys show promise of
raising that temperature, but on the other hand, the zirconium lowers the
fissile atom density close to that of carbide. As a result the second
generation of breeder reactors turned away from metal fuels.

PROTOTYPE REACTORS

About 1960 several countries began to think seriously about the next step in fast breeder reactor development. Several countries chose to take an inter- mediate step and build modest sized reactors with ceramic fuel cores, while also considering larger prototypes. Thus France built Rapsodie, a 20 MWt sodium-cooled reactor, mainly to test uranium-plutonium dioxide cores, while Russia built BOR-60 (or BOR), a 60 MWt sodium-cooled reactor, to test oxide and carbide cores. The USA built SEFOR, a 20 MWt sodium-cooled reactor with a mixed oxide core primarily for physics and safety experiments. Germany adapted the KNK reactor at Karlsruhe to operation as a fast reactor and Japan built JOYO, a 100 MWt reactor. However, the main interest has been in designing and building reactors of ~250-300 MWe size. Britain built the Prototype Fast Reactor (PFR) of 250 MWe capacity, France built Phenix (250 MWe) and the USSR built BN 350 (350 MWe). All are operating or have operated. The USA chose to build first FFTF (Fast Flux Test Facility) a 400 MWt test reactor at Richland — this is now operating. Germany, in association with several other countries, is building SNR-300 (300 MWe), while Russia and France are building BN-600 and Superphenix (1000 MWe), respectively.

Before discussing these cores further, we should address the philosophy behind the choice of fuel element design. The general aim of fast reactor designers around 1960 was to build reactors in which to develop potentially commercial fuel elements, power generation being a secondary aim, but even so it was more for the testing of major components than the generation of electricity *per se*. Given a desire to build a reactor whose core was reliable and of high performance, considerable attention was given (for a while) to cermet fuels. Cermets are dispersions of ceramic fuel in a metallic matrix. Thus each ceramic particle is surrounded by its own shell of cladding — rather like the HTGR fuel particles. Prior to 1960, cermets contained 30 v/o or less of fuel. For fast reactor use it was necessary to raise that level to ~50 v/o or higher and at the same time to ensure a uniform dispersion of the fuel particles. Good progress was made towards this (Fig. 10.10) [12] and some successful irradiations were carried out in the DFR core. However, it was soon realized that cermets absorbed too many neutrons and would never be considered for commercial cores. Of the alternatives it quickly became clear that oxide fuel was attractive, even though its breeding characteristics were inferior to carbide and metal fuels.

In 1961 some information was published on some high burn-up experiments on oxide fuels by GE in the GETR, which showed lower swelling and higher gas release than metal fuels [13]. This was backed up by the increasing avail- ability of data on UO_2 resulting from the US naval reactors programme [14]. It was known that plutonium formed solid solutions up to high concentrations in UO_2 (up to about 50% Pu in practice). Thus most countries decided, semi- independently, to opt for $(U,Pu)O_2$ as the first choice for an LMFBR core, with UO_2 as the axial and radial blanket material. It was realized that $(U,Pu)C$ had better breeding properties (Table 10.1), but would take longer to develop.

However, in the early to mid-1960s other decisions were being faced. Calcu- lations showed that a core rated at about 30 kW/m would be about 1 m high by 1.75 m diameter with a fuel pin diameter of about 0.575 cm (0.23 in). The latter was determined by the need to avoid fuel centre melting. Given the small pin diameter and 1 m active core length, it was decided to aggregate the pins into a hexagonal subassembly of about 200 pins, spaced apart by wires or grids.

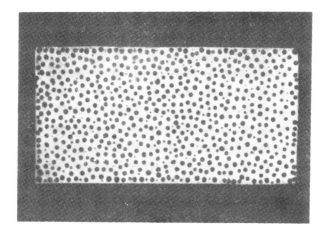

Fig. 10.10.   UO$_2$-stainless steel cermet. Near-ideal
particle distribution. Source: Frost and
Waldron.

An important decision was whether to use the fuel in the "Vipac" form or as
pellets. Vipac had been developed by the Hanford Laboratory for a thermal
reactor fuel recycle project (Pheonix). Instead of pressing and sintering
fuel precisely into pellets, it was roughly pressed and sintered then broken
up and sized by sieving into three size fractions which would pack down in a
tube to a fairly high density. Alternatively, spheres could be made by a
sol-gel process in the required sizes and packed in a mechanically vibrated
tube (hence "Vipac"). This packed bed has an even lower thermal conductivity
than pellets, with consequently higher centre temperatures, so it was decided
in the USA to work with pellets. The UK, believing that some voidage for
swelling was needed, developed annular pellets (like cotton reels). However,
as restructured fuels tell us — a central void does not help much in accom-
modating swelling.* End-dishing of pellets appears to be more useful for
this purpose and is generally used. One should note parenthetically that
the sol-gel process is enjoying a revival because limited experience with
remote fabrication in shielded gloveboxes has shown dust to be a particular
problem in maintaining or replacing equipment and sol-gel fuel creates very
little dust.

Another decision that had to be made was the choice of cladding. Physics
considerations drove this towards stainless steel, although Argonne and
Westinghouse developed vanadium alloys as an alternative [15]; these were
unable to overcome the problem of embrittlement by residual carbon and oxygen
in the sodium. The properties of Type 316 stainless steel appeared to sur-
pass those of its rivals and it was selected first by the UK and later by the
US, USSR, France and others. Radiation damage data at that time showed that
loss of ductility was likely to be a key problem. Calculations to derive the
cladding wall thickness assumed that the maximum stress should never exceed
the yield stress (which increases with radiation dose) and the stress should
not exceed that which would cause 0.1% creep in 10^4 hours. While good
elevated temperature creep data were available for Type 316 stainless steel,

---

*Nevertheless, annular fuels appear to be making a comeback in LWRs.

the effects of irradiation were unknown other than the embrittling effect (which led to the 0.1% creep limit). Plenums were obviously necessary to bring gas pressures down to a level where the wall thickness was not excessive. A plenum of length close to that of the fuel column was chosen and this gave a wall thickness of about 0.375 mm (0.015 in). Since this is a thin wall it was specified that there should be at least 10 grains across the wall and that nonmetallic inclusions should be absent. This led to the adoption of double vacuum arc melting to produce the blanks for tube drawing. The allowable flaw size was less than 10% of the wall thickness — a size which was close to the limits of NDE processes.

Prior to settling on thin-walled stainless steel cladding combined with gas plenums, consideration was given to thick, highly restraining cladding as a means of attaining high burn-up; the trade-off between higher burn-up and higher neutron absorption was debated. Blake [16] proposed using unalloyed uranium, UO_2, or even a cermet in a strong can to attain high burn-up. He obviously was unaware of the phenomenon that led to the EBR-II Mk II fuel pin design — the breakaway gas release at 25% swelling, which offers an alternative. Zebroski [17] and his colleagues at GE carried out valuable experiments on highly restrained UO_2 and $(U,Pu)O_2$; often referred to as gun-barrel experiments. Eventually the physicists won and cladding thickness was reduced to a minimum level and this decision has proven to be the right one.

By about 1960 oxide fuel fabrication lines had been established on a scale consistent with making small experimental subassemblies for irradiation in the EBR-II and DFR reactors and there followed an extensive series of irradiations. These were initially parametric in nature, i.e. the pin designs were varied with respect to fuel density, fuel-clad gap, fuel : plenum ratio, cladding type (e.g. degree of cold work), wire-wrap versus grids, etc. Until these parametric tests had been run and evaluated it was not possible to begin to firm up on a fuel element design with confidence.

In the course of this testing some unexpected phenomena made themselves apparent. The most important was the discovery of radiation-induced voids in 1967 when Dounreay workers examined stainless steel that had been exposed in the DFR core [18]. It took a while for the implications of this discovery to be fully realized, but when they were large programmes were started to develop low-swelling alloys. Some simple remedies, like adding more silicon and titanium to Type 316 stainless steel, were fairly successful (US designation of this alloy is D9). Some commercial alloys, such as the British PE16, gave lower swelling, and the ferritic stainless steels gave even lower swelling. In addition, redesign of cores was undertaken to accommodate the effects of swelling of fuel cladding and, more particularly, the subassembly ducts.

Another unexpected result was the incidence of cladding attack by certain fission products, described in Chapter 3. The effect was discovered by metallographic observation of attack on the cladding inner wall and chemical studies discovered the mechanism [19]. Cures were proposed in the form of lowering the fuel O/M ratio and of coating the cladding inner wall with a reactive metal such as titanium or chromium; these have been tested in-reactor. Lowering the O/M ratio causes a secondary effect of caesium migrating to the UO_2 axial blanket pellet where it forms caesium uranate, expanding the fuel column locally and leading to cladding failure. Hence the O/M adjustment must be precise, or the coating or gettering approach may be needed.

Having discussed the design questions, let us now discuss what has evolved
as a design. Figure 10.11 shows the FFTF fuel pin design and its subassembly,
which is typical of worldwide designs. The $(U,Pu)O_2$ fuel pellets are pressed,
sintered and centreless ground to a diameter of 0.200 in and a length of
0.2 in, with a dimple at each end to allow for pellet expansion. The pellets
are loaded into 0.200 in (0.51 cm) i.d., 0.230 in (0.58 cm) o.d. 20% cold-
worked Type 316 stainless steel cladding which has a spiral wire wrap spacer
welded to the outside. The pellet density is ∼94% and combined with the fuel-
cladding gap and the dimples gives an overall of "smear density" of about 88%
of theoretical density (these values may vary but are typical). The Pu
content is∼20% and the O/M ratio 1.94. The fuel column length is 36 in.   In
FFTF there is a nickel reflector section above and below the fuel.   In a
power reactor there would be ∼18 in (46 cm) of stoichiometric UO_2 blanket
pellets in that position. There is a spring that holds down the pellet stack
to prevent movement. A 42 in (106 cm) gas plenum occupies the upper end of
the element. In the PFR a bottom plenum is used, in part because it is
exposed to the cooler inlet sodium and in part because its diameter is
reduced to permit improved flow to the fuel section. In Fig. 10.11 one can
see a tag gas capsule in the plenum. Should the element fail, this releases
a predetermined mixture of xenon isotopes into the coolant. These give the
operators a characteristic signal on their coolant monitors, allowing them
to identify the location of the failure.

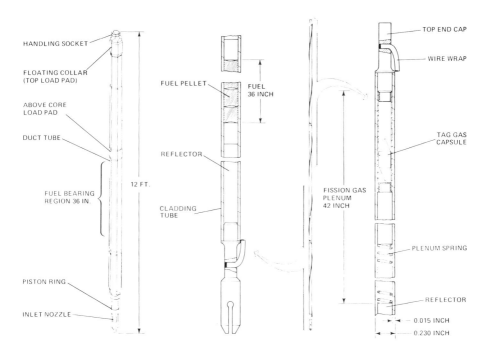

Fig. 10.11.   FFTF fuel element and subassembly. Courtesy:
Hanford Engineering Development Laboratory.

The FFTF subassembly is 4.575 in (12 cm) across the flats and holds 217 pins.
The PFR assembly is 5.7 in (14.4 cm) across the flats and holds 325 pins.
Both assemblies are about 12 ft ($\sim$3.6 m) in length and generate $\sim$6 MW of heat.

Typical performance parameters for current mixed oxide fuel pins in peak core
positions are:

rating	$\sim$200 W/g
	$\sim$40 kW/m
maximum clad temperature	$\sim$660°C
peak burn-up	12 at %
peak fluence at cladding	$\sim$10^{23} nvt

Extensive testing of mixed oxide subassemblies has been performed in DFR,
EBR-II, Rapsodie, Phenix and PFR. In general, the failure rate has been low
and generally in the form of "infant mortality" due to errors in welding or
assembly. Burn-ups in excess of 100,000 MWD/te or $\sim$10% heavy atoms have
attained regularly [20], giving confidence in future performance. However,
one must add a note of caution. In order to test fuel elements in the smaller
reactors (EBR-II, DFR and Rapsodie) the fuel has to contain fully enriched
uranium. This alters the chemistry of the fuel, because most fissions occur
in uranium whose fission product yield is different from that of plutonium.
Second, while one can attain the desired fuel rating by this means, the fast
neutron flux is lower by an order of magnitude than that in a large reactor,
so that the neutron fluence : burn-up ratio in these tests is not typical of
a large reactor. The cladding will behave better because it accumulates
less damage per increment of fuel burn-up. The larger reactors like FFTF,
Phenix and PFR do not have this disadvantage.

Having discussed the "reference" oxide fuel for fast reactors let us discuss
alternatives for improved performance. We have already discussed metallic
fuels and noted their advantages and disadvantages. That leaves carbides and
nitrides, which are very similar to one another in properties and performance.
Nitride suffers disadvantages in that it requires enrichment in ^{15}N in order
to compete with carbide neutronically and it does not melt at high temperatures — it vaporizes — and the consequences of this on safety are not clear.
Thus we will concentrate our discussion on carbide as an alternative to
oxides.

Although three carbides of uranium are known to exist, it is the monocarbide
UC or (U,Pu)C in which we are interested; partly because it has the least
number of carbon atoms per "molecule" and partly because it has high
stability. The advantages of carbide over oxide fuel for a fast reactor
core are:

1. Higher heavy metal density (13.0 versus 9.7 g/cm^3), hence better breeding
    (see Table 10.1).
2. Higher thermal conductivity.
3. One moderating atom (C) versus two (O$_2$).

One may use the higher thermal conductivity to produce larger diameter pins.
and hence a smaller number of them. This is desirable because the cost of
fabricating carbide is about 25% higher than oxide because of the extra
steps involved.

Some disadvantages of carbide over oxide are:

1. Brittleness — leading to easy fracture.
2. Retention of fission gases.
3. Invariant composition.

Unlike UO_2, UC cannot exist over a composition range. A little less carbon produces free uranium with bad effects on performance; a little more carbon produces UC_2 or U_2C_3, which tends to react with steel cladding. A little excess U_2C_3 seems to be acceptable, and generally this is the structure which is aimed at.

With carbide (and nitride) fuels we have two options — a sodium bond with a large fuel-clad gap, or a helium bond. Testing has been carried out on both bonds. Figure 10.12 plots the fuel centre temperatures as a function of specific power for sodium bonded and helium-bonded carbide fuel and for oxide with a helium bond. The sodium-bonded carbide clearly gives the lowest fuel temperature. However, there is a tendency for the fuel to crack into large chunks in the sodium-bonded version (Fig. 10.13 shows this for a nitride fuel, but carbide behaves similarly). The chunks can form a "log-jam" that stresses the cladding locally leading to failure. Crushable liners have been placed in the bond to avoid this effect, but one begins to get a complicated design [21].

The swelling of UC and UN increases with fuel temperature as shown in Fig. 10.14, hence it is desirable to stay below $\sim$1400°C especially in a helium-bonded design. However, at 1400°C and below, the gas release is low and fission gases reside on the grain boundaries (Fig. 10.15) [22]. Sudden changes in power or temperature may release these gases through bubble linkage, or it may result in sudden gross swelling or "bloating". So it appears to be difficult to find a good compromise for carbide fuel pin design. However, a number of helium-bonded pins have been irradiated to high burn-ups without failure [21] and transient tests have not revealed any unusual or alarming effects.

The process of fuel element parametric testing has interacted strongly with the computer modelling activities. Generally the models were used to predict the outcome of the tests and the results were compared with the post-irradiation results; discrepancies led to refinement of the codes. However, the code predictions are only as good as the materials properties data used in the codes and there is room for improvement in this area. Property measurements have become an unpopular activity, but must be revived if codes are to achieve their full predictive potential. Much of the discussion on modelling in Chapter 5 relates to fast reactor fuel pin modelling. The LIFE and UNCLE codes were originally developed for fast reactor fuel pins.

As the design of the fuel element has become clearer and more generally accepted, safety questions have assumed greater importance. These fall into the categories of minor operational problems, where the main issue is whether the core can continue to operate after a minor transient event, and major accident conditions which might lead to a hazard to the population. In attacking both kinds of problem a similar methodology is used; models are developed to predict the course of events and experiments are performed in special facilities to simulate accident conditions.

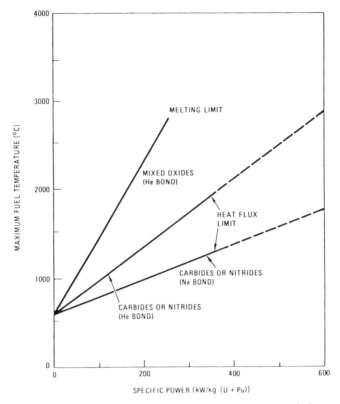

Fig. 10.12.   Fuel temperature as affected by specific
power (0.64 cm diameter pin). <u>Source:</u>
M. T. Simnad and J. P. Howe, Materials for
Nuclear Fission Power Reactor Technology.

Operational questions, that relate to the limits to fuel pin performance,
are operation with failed cladding (usually called Run-Beyond-Cladding-Breach
or RBCB) and mild overpower transients ∿130% full power. The main problems
in RBCB are the spread of activity around the reactor primary circuit
(particularly the entrainment of powdered fuel) and the rate of reaction of
the sodium coolant with the fuel to form a low density compound, leading to
more gross rupture of the cladding. Related to this latter question is the
problem of pin-to-pin failure propagation, i.e. when the cladding ruptures
does the efflux of fission gases or the constriction of the coolant channel
cause the adjacent pins to fail, leading to a cascade of failures?

RBCB and pin-to-pin failure propagation problems are tested in sodium loops,
such as the SLSF series of loops being run in the Engineering Test Reactor,
and in operating reactors. Once the FTR is operating it is proposed to use
the EBR-II reactor as a test vehicle for RBCB experiments.

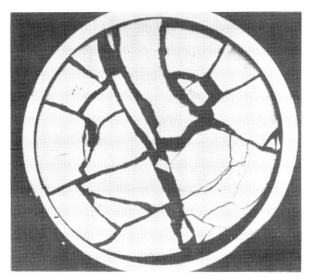

Sodium–bonded (20 mil radial),
high density fuel
Burnup–152,000 MWd / MTM

Fig. 10.13.   Fuel cracking in mixed nitride fuel pin.
Source: A. Bauer *et al.*, Fast Reactor Fuel
Element Technology, ANS, 1971.

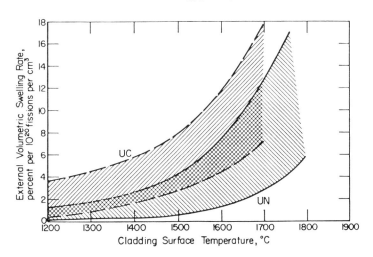

Fig. 10.14.   Comparison of volumetric swelling rate as a
function of cladding surface temperature for
UC and UN clad in W–25 wt% Re.   Source:
R. F. Hilbert *et al.*, Fast Reactor Fuel
Element Technology.

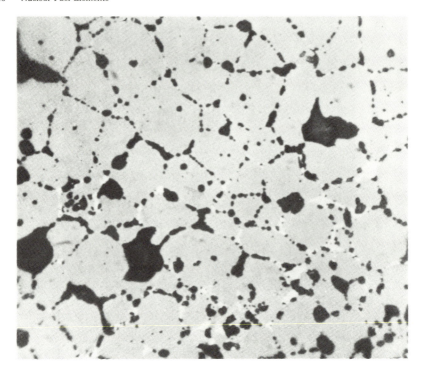

Fig. 10.15. Linking of grain boundary gas bubbles near
mid-radius of Cr-modified fuel (pin W8X),
500X. Source: B. Harbourne *et al.*, Reactor
Fuel Element Technology.

The potential initiators of major accidents in fast reactors are overpower
transients and loss of coolant flow. These are examined experimentally in
transient test reactors, of which TREAT in the USA and CABRI in France are
the only purpose-designed examples operating today. Typically in TREAT a
sodium loop containing a small cluster of EBR-II size fuel pins (most likely
pre-irradiated in EBR-II) is inserted in the reactor and exposed to a
predetermined performance pattern. For example, steady temperature conditions
are established in the fuel and then the TREAT core is pulsed to subject the
fuel to a predetermined power rise rate until it reaches a certain level and
is cut off. The time of the transient is typically 20 seconds. After the
"shot" the loop is removed from the reactor, is neutron radiographed and
disassembled to determine the nature of the fuel pin damage.

Such experiments in the course of LMFBR development showed the need to mix
the ^{238}U and ^{239}Pu atoms in such a way that heat flows from the plutonium to
the uranium very rapidly. This is because of the Doppler effect whereby the
rising fission rate of the Pu is balanced by the falling fission cross-section
of the U as its temperature rises. This had an effect on fuel specifications
and the choice of fuel forms. For example, it requires that no uranium-rich
or plutonium-rich particles greater than ~100 µm exist in the fuel.

The safety performances of the three main fuel types are expected to be different from one another. Because of the sodium bond design, metal fuels are expected to swell until their rate of gas release is high; the main safety question is that of reaction of fuel with the cladding. However, even if a fuel-cladding reaction occurs the product will drip down to the bottom of the core. Hence most metal-fuelled fast reactors have had flow separators at the base of the core to prevent recriticality.

Oxide fuels have high melting points, but generally release their fission gas. Hence in an accident the cladding will melt first and a key issue is the stability of the unsupported fuel stack. This is being attacked both in TREAT tests and by direct electrical heating (DEH) of irradiated fuel stacks for some time, during which molten fuel is ejected through cracks in the pellets (UO_2 expands by about 10% on melting). Carbides may retain a lot of fission gas that is released in a transient.

Another facet of accident analysis is fuel-coolant interactions. If a core overheats and vaporizes the sodium which subsequently condenses and re-enters the core, there may be an explosive reaction between the re-entering sodium and the now very hot fuel. Because of its different physical properties, carbide may produce a larger fuel-coolant interaction than oxide.

Many of these questions are unanswered at this time, particularly with respect to the behaviour of full-size subassemblies. To study subassemblies, TREAT is being modified and a Safety Test Reactor has been designed. However, one should not be discouraged by this lack of information because a sodium-cooled LMFBR is inherently safer in many ways than an LWR because the low-pressure sodium can carry away heat by thermal convection. Thus a pipe-break does not lead to coolant boiling and the heat removal capacity of naturally convecting sodium is more than enough to remove the decay heat. Most of the safety efforts on LMFBRs at this time are devoted to studying highly improbable events, and indeed to place a probability on those events. As breeder reactor fuel development has progressed there has been a change in emphasis from steady state performance to accident performance as the reactor type approaches commercialization and licensing.

Since 1977 the LMFBR has been subject to special scrutiny with respect to nuclear weapon proliferation. The fuel contains approxiamtely 20% of plutonium, and might just be usable for a weapon. In the fuel cycle plutonium is separated, both from the spent LWR fuels and from the irradiated LMFBR fuel. This would be a very attractive source of weapon material. Hence, there has been a lot of study of proliferation-resistant systems. These range from banning reprocessing, through once-through breeders such as the Fast Mixed Spectrum Reactor (see later) to schemes in which high gamma activity is retained in the fuel to prevent its diversion. The lesson here is that non-technical considerations may influence design strategy and these are usually difficult to anticipate.

Despite all of the questions raised in the course of development, two 300 MWe LMFBRs are currently operating satisfactorily with mixed oxide cores: Phenix and PFR. FFTR has gone to power. The Superphenix, SNR-300 and Monju are under construction and will use mixed oxide cores for start-up. Experience with stainless-steel clad mixed oxide fuel elements in Rapsodie and Phenix cores has been good, with a fairly low failure rate up to 10 a/o burn-up of the fuel. Criteria and strategies for the introduction of advanced cores have not yet been worked out because the necessary data on performance and safety are lacking.

The development of the LMFBR has spawned two variations that deserve comment. US and European experience in the development of gas-cooled reactors, especially the HTGR, led to consideration of a gas-cooled fast reactor (GCFR)[23]. This might lead to greater thermal efficiency, better breeding (because of low neutron capture in the coolant) and good safety. Hence, the best features of the HTGR and the LMFBR were combined in the GCFR. Pre-stressed concrete pressure vessel technology was used to provide a very safe structure and coolant-handling system (Fig. 10.16). The core used the LMFBR fuel and cladding technology. However, the heat transfer from the stainless steel cladding is much poorer to 1000 psi helium than it is to 100 psi sodium. To overcome this problem the cladding was finned to promote turbu-lence in the coolant boundary layer. Another problem was the high coolant pressure which would cause the cladding to collapse onto the fuel and "breathe" as the reactor power varied. To avoid this happening the fuel pin was vented to the coolant, i.e. the pressures inside and outside the pin were equalized. This then required that the fission products be trapped or isolated by charcoal or metal traps, following HTGR technology. The venting concept has been tested in thermal reactor loops and the roughening concept has been tested in heat transfer loops. Only limited testing of fuel bundles in a fast flux has been carried out. The availability of loops to test full-size subassemblies in the FTR could carry the concept forward to the stage where a demonstration plant can be built with confidence.

The political problem of proliferation-resistance led to a revival of an old concept, the Fast Mixed Spectrum Reactor [24]. This, as the name implies, is a reactor with both fast and thermal zones. At equilibrium fuel is placed in the outer thermal zone where it builds up plutonium. It is moved in steps to the central fast zone where the plutonium is burned. It is then moved out to the thermal zone again to burn further fissile atoms. The fuel need not be reprocessed since the fuel is natural or low-enrichment uranium. The physics of this reactor is delicately balanced and metal fuel is needed to make it work. The metal fuel must sustain a burn-up of ~ 12–14% and the cladding will experience a neutron fluence approaching 10^{24} n/cm^2. The burn-up should be attainable using the EBR-II Mk II design, but evaluation of the cladding will be very difficult because it would take many years to approach the flux in existing fast reactors. A possible approach is to take the highest dose material available at this time ($\sim 2 \times 10^{23}$ n cm^{-2}) and subject this to heavy ion bombardment to determine whether any new phenomena occur at very high doses. The basic fuel pin design is likely to resemble the EBR-II Mk II design, i.e. metal fuel with a wide sodium bond and a sealed plenum so that EBR-II experience will be very relevant.

Finally we should mention two aspects of the behaviour of fuel subassemblies in fast reactors. First, the subassemblies are located in a tight array in the reactor core and blanket. Various means are used to ensure the continued tightness of the core array during its lifetime. One is shown in Fig. 10.17. In the FFTF, and other designs, a peripheral core restraint mechanism applies a force to two peripheral reflector assemblies at two elevations above the active core. One has to remember that neutron damage will be causing differential swelling of the subassemblies which will be working against the restraint. This may give rise to radiation-induced creep or stress-relaxation, so that the final shape of the subassembly is difficult to predict. A movable restraint mechanism will permit easier unloading of a warped core.

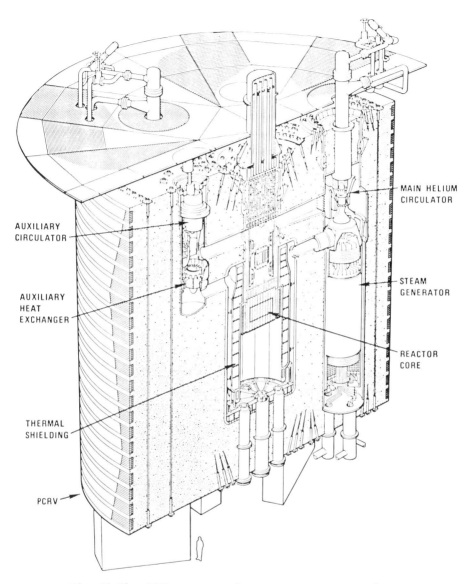

AUXILIARY
CIRCULATOR

AUXILIARY
HEAT
EXCHANGER

THERMAL
SHIELDING

PCRV

MAIN HELIUM
CIRCULATOR

STEAM
GENERATOR

REACTOR
CORE

Fig. 10.16.   GCFR prestressed concrete reactor vessel
arrangement. Courtesy: General Atomic
Company.
The primary system of a GCFR is contained within a large
concrete vessel. Various components are located in
cavities within the vessel, and penetrations into the
vessel are used for servicing these components.

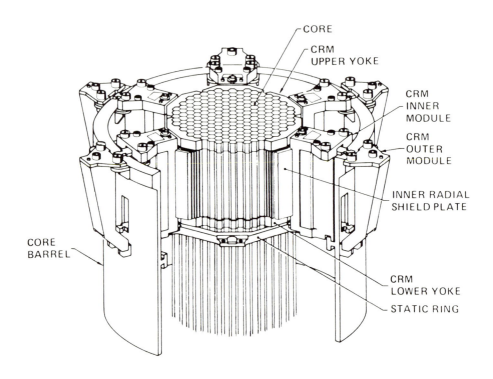

Fig. 10.17. Core restraint on FFTF. <u>Courtesy</u>:
Westinghouse Electric Corporation.

Refuelling of fast reactors is not done in the same way as for thermal reactors. Figure 10.18 shows the most common arrangement for large fast reactors. For obvious reasons it is known as an A-frame. The fuel transfer lock swings the subassemblies from the reactor head over to the chute that leads to the storage pool where the subassemblies may cool for at least 120 days before transfer to the reprocessing plant. Fresh fuel is held in a storage carousel outside the main reactor hall ready for handling by the transfer lock (Fig. 10.18 item 8).

Refueling system

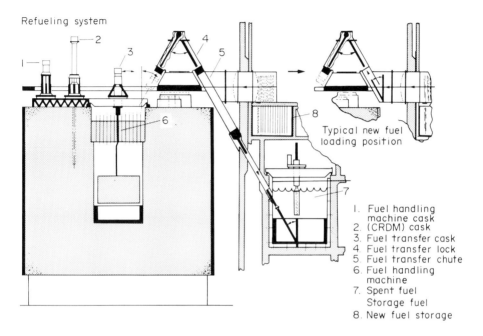

Typical new fuel loading position

1. Fuel handling machine cask
2. (CRDM) cask
3. Fuel transfer cask
4. Fuel transfer lock
5. Fuel transfer chute
6. Fuel handling machine
7. Spent fuel Storage fuel
8. New fuel storage

Fig. 10.18.   Typical fast reactor refuelling system.
Courtesy: General Atomic Corporation.

There is controversy over the timescale on which fast reactors are needed. It depends on the rate of increase in electricity demand, the size of the uranium reserves and the rate of development of fusion power. However, most developed countries believe that breeder reactors will be needed for commercial generation of electricity before the end of the 20th century. This means that the fuel development process must advance to the licensability stage by 1990. Fortunately it seems that this process is well along and this goal should be achievable.

In following the stages of development of fast reactor fuel elements through to today, the reader is reminded to turn back to Table 1.1 which traces the many stages of fuel element development.

REFERENCES

1. Kittel, J. H., Argonne National Laboratory Report ANL-5731 (November 1957).
2. Leipunskii, A. I., *et al.*, *Proceedings of the 3rd International Conference on the Peaceful Uses of Atomic Energy*, Geneva, Paper P/312, UN (1964).
3. Kazachkowski, O. D., *et al.*, *At. Energ. (USSR)* 24, 136 (1968).
4. McIntosh, A. B. and Bagley, K. Q., *J. Brit. Nucl. Energy Conf.* 3, 15 (1958).
5. Symposium on the Dounreay Fast Reactor, 7th Dec. 1960, Institute of Mechanical Engineers, and British Nuclear Energy Conference, London.
6. Swanson, K. M., Sloss, W. and Batey, W., *Proceedings of Topical Meeting on Advanced LMFBR Fuels*, Tucson, AZ (October 1977), American Nuclear Society (1977).
7. McDaniel, W. N., *et al.*, *Proceedings of the 2nd International Conference on the Peaceful Uses of Atomic Energy*, Geneva, 1958, Paper P/792, UN (1958).
8. Raiteck, R. C., *et al.*, *Trans. Am. Nucl. Soc.* 11, 100 (June 1968).
9. Koch, L. J., *et al.*, *Proceedings of the 3rd International Conference on the Peaceful Uses of Atomic Energy*, Geneva, Paper P/207, UN (1964).
10. Hesson, J. C., Feldman, M. J. and Burris, L., Argonne National Laboratory Report ANL-6605 (April 1963).
11. Walter, C. M., Golden, G. H. and Olson, N. J., Argonne National Laboratory Report ANL-76-28 (November 1975).
12. Frost, B. R. T., *et al.*, *Proceedings of the 3rd International Conference on the Peaceful Uses of Atomic Energy*, Geneva, 1964, Paper P/153, UN (1964).
13. Gerhart, J. M., *et al.*, USAEC Report GEAP-3833 (1961).
14. Daniel, R. C., Bleiberg, M. L., Meieran, M. B. and Yeniscavich, W. B., USAEC Report WAPD-263 (1962).
15. Edison, G., *et al.*, WARD-3791-19 (July 1968).
16. Blake, L. R., *Reac. Sci. and Tech. (JNE* Parts A/B) 14, 31 (1961).
17. Nelson, R. C. and Zebroski, E. L., *Trans. Am. Nucl. Soc.* 9, 412 (Oct-Nov 1966).
18. Cawthorne, C. and Fulton, E. J., *Nature*, 216, 575 (1967).
19. Johnson, C. E., Johnson, I. and Crouthamel, C. E., *Proceedings of Conference on Fast Reactor Fuel Element Technology*, New Orleans, LA, 1971, p. 393, American Nuclear Society (1971).
20. *Proceedings of a Conference on Advanced LMFBR Fuels*, Tucson, AZ, Oct. 10-13, 1977, American Nuclear Society (1977).
21. Barner, J. O., *et al.*, ibid, p. 268.
22. Harbourne, B. L., Levine, P. J., Latimer, T. W. and Neimark, L. A., *Proceedings of Conference on Fast Reactor Fuel Performance*, New Orleans, LA (1971), American Nuclear Society, p. 869.
23. *Nucl. Eng. Design*, Special GCFR Issue, 40 (1), 1-233 (January 1977).
24. Fischer, G. J. and Cerbone, R. J., BNL-50976 (January 1976). Obtainable from NTIS.

CHAPTER 11

Research and Test Reactor Fuel Elements

Most of this book is concerned with fuel elements for power reactors. For completeness we must discuss fuel elements for research and test reactors. Research reactors are fission neutron sources that are used for basic research. This includes radiation effects on materials and on biological samples, neutron scattering studies of crystal structures and atomic properties, and the production of isotopes for research purposes. Test reactors are generally built to carry out technological studies, such as the testing of fuel elements in loops and the exposure of fuel elements to power or coolant transients.

Research needs have pushed research reactors to higher fluxes and powers so that the distinction between research and test reactors has become blurred. Some reactors fulfill multiple purposes, providing beams of neutrons for research, loop or capsule tests for technological purposes, and isotope production and even the irradiation of plants and animals.

The majority of the world's research and test reactors are water cooled and moderated. Since they are not required to generate useful power, the water may be kept as cool as practical and preferably is not pressurized. The swimming pool type of reactor is the simplest example. However, there has been a trend towards higher and higher neutron fluxes. Fluxes are proportional to the fission rate or power density. This then puts a premium on heat transfer. Indeed, the limit to core power density is the departure from nucleate boiling (DNB) or dryout, which leads to rapid overheating of fuel and its failure or rupture. Hence, the trend in research and test reactor fuel element design has been towards maximizing the surface area to volume ratio and to pressurization of the coolant/moderator.

In pursuit of a high power density without large power generation these reactors have been designed with compact cores containing fully enriched uranium which is burned up rapidly. In the interests of high reactor availability the fuel must be capable of sustaining a high burn-up without failure.

The landmark in research reactors was the Materials Testing Reactor (MTR) which was a 40 MW light-water-cooled and moderated research reactor that first achieved full power in May 1952. The landmark was the fuel element design [1]. The fuel was an alloy of 18 w/o fully enriched uranium and 82 w/o aluminium in plate form and clad in aluminium. The fuel was made by vacuum melting and casting into slabs which were hot rolled to strip, from which cores were punched. The alloy consisted of a dispersion of particles of UAl_3 in an aluminium matrix. Fission events are thus isolated to the UAl_3 particles each of which is surrounded by a load-bearing, high thermal conductivity web of aluminium. This provides a low operating temperature and low swelling (see later).

The fuel cores were fabricated into fuel plates by the "picture frame" technique (Fig. 11.1). The core was fitted into a frame on top and bottom of which plates were fitted and edge-welded. The assembly was outgassed through an evacuation tube which was crimped shut. It was then rolled at 590^0C to a total reduction of 84% (Fig. 11.2). The plates could then be shaped in a die. In the MTR case (and its successors) the plates are curved and are then mounted in a box-shaped assembly and brazed into the final shape (Fig. 11.3). The curvature of the plates was designed to provide stiffness and resistance to buckling. The fuel element box is fitted with end pieces that aid alignment and flow in the core.

The original MTR fuel assemblies contained 140 g of ^{235}U, later increased to 200 g. These gave an average burn-up of ~25% of fissile atoms in a core residence time of 3 weeks. In this configuration the peak thermal neutron flux was 9×10^{14} $n/cm^2/sec$ and the average was 2 to 3×10^{14} $n/cm^2/sec$. Few failures were experienced; those that were observed were a consequence of design changes that led to buckling of the plates.

The cores of this type of plate element could be made by powder metallurgy instead of melting and casting. Aluminium powder of suitable size was mixed with crushed UAl_3, that had been made by melting and casting, and pressed in a rectangular die. This gave a more uniform and controlled fissile particle size and distribution than the cast material.

The MTR fuel element design was so successful that it was used in a number of other research reactors, especially the Oak Ridge Research Reactor (ORR) which has been reproduced at Studsvik in Sweden (R2 reactor) and elsewhere. The most advanced application of this type of element was in the High Flux Beam Reactor at Brookhaven [2]. This is a 40 MWt reactor with a peak thermal flux of 7×10^{14} $n/cm^2/sec$. The peak surface heat flux is 1.6×10^6 $Btu/hr\ ft^2$ under a water pressure of 250 psig and temperature of 150^0F which is fairly close to DNB conditions.

The HFBR fuel element (Fig. 11.4) contains a core of 30 wt % uranium (at 93% enrichment), 3 wt % silicon and 67 wt % aluminium. The silicon promotes the formation of UAl_3 in preference to UAl_4 and gives a more uniform alloy. The ^{235}U content of each element is 274 g and a full core loading contains 7.67 kg of ^{235}U. The fuel-plate frame and cladding material are made of aluminium alloy 6061 which contains 0.8-1.2 wt % Mg, 0.4-0.8 wt % Si, 0.7 wt % Fe, 0.15-0.40 wt % Cu, 0.15 wt % Mn, 0.25 wt % Cr, 0.25 wt % Zn; balance is Al. The fuel element life is 38 days with an average burn-up of 20% ^{235}U.

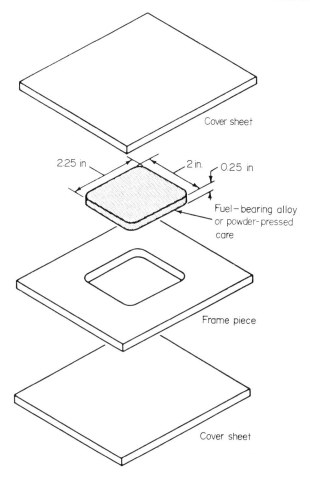

Fig. 11.1. Picture frame technique for fuel plate
fabrication. <u>Courtesy</u>: Oak Ridge National
Laboratory.

The Engineering Test Reactor (ETR) is a light-water moderated and cooled loop
test reactor of 175 MWt power, operated at INEL in Idaho. The fuel element
is again similar to the MTR design with 19 plates per assembly. The full
core loading contains 49 assemblies measuring 3.0 x 3.0 x 55 in in a
10 x 10 array with a ^{235}U content of 13 kg. The normal lifetime is 3 weeks
at full power to a burn-up of ∿32%. The average thermal flux is
4 x 10^{14} n/cm^2/sec.

A variation on the MTR plate design is the concentric fuel plate design
typified by the Argonne CP-5 reactor. The CP-5 element contained three
concentric fuel tubes which gave improved rigidity, greater coolant flow and
accessibility for fast flux irradiations in the centre of fuel assemblies
(Fig. 11.5). The fuel tubes were made by an extrusion process [3]. First

cast U-Al alloy rods were bored out and extruded into fuel tubes. These
were placed in aluminium 6061 "frames", welded up and extruded to final shape.
These tubes were then assembled into the final configuration. Since the heavy-
water moderated and cooled reactor only operated at 2-5 MW, the burn-up rate
of the elements was low. A typical core residence time was 2 years with a
peak burn-up of 45% and an average of ~25%.

The Belgian BR-2 reactor is a 50 MW light-water-cooled and moderated research
and test reactor [4]. The thermal flux is close to 10^{15} n/cm^2/sec. The core
is contained in an aluminium pressure vessel (like the ORR) inside a pool.
The normal coolant pressure is 180 psig and the temperature about 200°F.

Each fuel assembly contains six concentric tubes similar to the CP-5 tubes —
a U-Al "picture frame" type. As seen in Fig. 11.6, the tubes are spaced
accurately by means of ribs or spacers. Each assembly contains 250 g of
^{235}U and the average critical mass of the core is ~4 kg of ^{235}U.

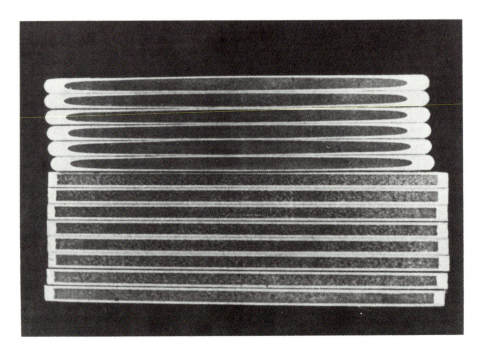

Fig. 11.2. Cross-section of fuel plates containing meat
            consisting of a U$_3$O$_8$ Al-cermet
            (Above) Fabricated by coextrusion: elliptical
                    cross-section, nonuniform can thickness.
            (Below) Fabricated by the picture-frame
                    technique: rectangular cross-section,
                    uniform thickness.
            Source: NUKEM.

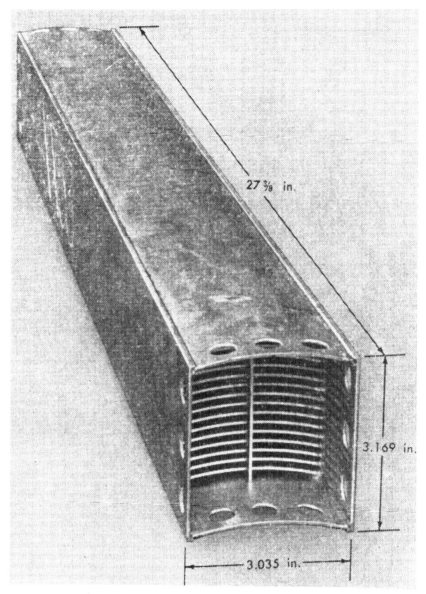

Fig. 11.3. MTR fuel box. <u>Courtesy</u>: Oak Ridge National
Laboratory.

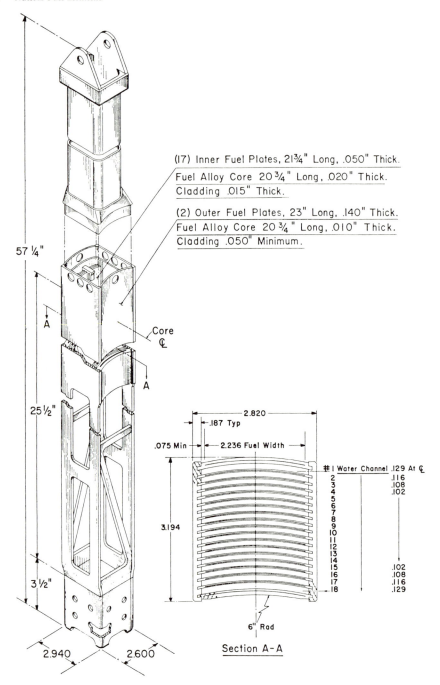

(17) Inner Fuel Plates, 21¾" Long, .050" Thick.
Fuel Alloy Core 20¾" Long, .020" Thick.
Cladding .015" Thick.

(2) Outer Fuel Plates, 23" Long, .140" Thick.
Fuel Alloy Core 20¾" Long, .010" Thick.
Cladding .050" Minimum.

Fig. 11.4.   The HFBR fuel element.   Courtesy: Brookhaven
National Laboratory.

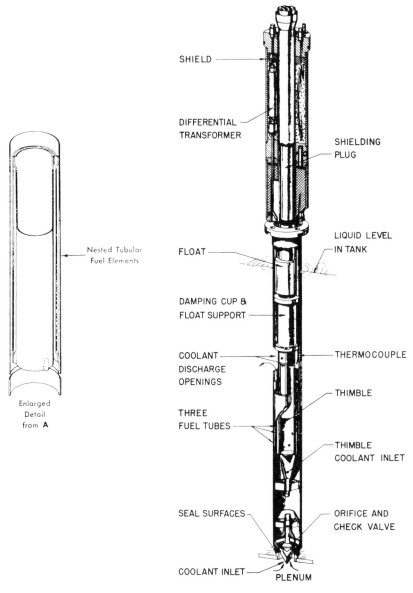

SHIELD

DIFFERENTIAL TRANSFORMER

SHIELDING PLUG

Nested Tubular Fuel Elements

FLOAT

LIQUID LEVEL IN TANK

DAMPING CUP & FLOAT SUPPORT

COOLANT DISCHARGE OPENINGS

THERMOCOUPLE

THIMBLE

Enlarged Detail from **A**

THREE FUEL TUBES

THIMBLE COOLANT INLET

SEAL SURFACES

ORIFICE AND CHECK VALVE

COOLANT INLET

PLENUM

Fig. 11.5.   Nested tubular fuel elements in operation in
             the CP-5 reactor. (Modified from an Argonne
             National Laboratory drawing.)

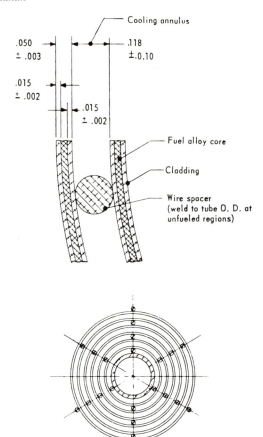

Fig. 11.6. BR-2 fuel element cross section. <u>Source:</u> Dopchie & Planquart, 2nd ICPUAE, 1958.

The largest test reactor in the USA (and probably the world) is the 250 MW Advanced Test Reactor (ATR) which operates at INEL [5]. The core is contained in a pressurized steel tank. The reactor was designed for complex loop experiments at thermal neutron fluxes of 1.5×10^{15} n/cm^2/sec at the experiment. The core layout is unusual (Fig. 11.7), in that the fuel elements follow a serpentine path around the experiments. Each fuel element (Fig. 11.8) is made up of curved plates set into side plates which are at an angle of $45°$ to one another. The fuel is a dispersion of U_3O_8 in an X8001 aluminium alloy with B_4C burnable poison added. The plates are roll clad, like MTR plates, with 6061 aluminium alloy [6]. A full core loading is 35-39 kg of ^{235}U. The coolant pressure is 355 psig and the maximum power density is 2.8 MW/litre, which is very high and close to the limit for DNB. The normal operating cycle is 17 days.

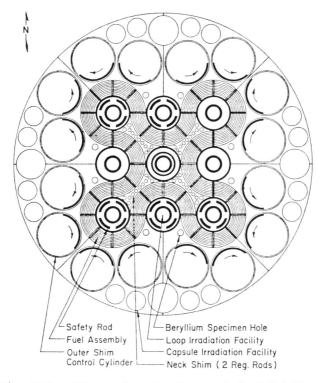

N

Safety Rod
Fuel Assembly
Outer Shim
Control Cylinder

Beryllium Specimen Hole
Loop Irradiation Facility
Capsule Irradiation Facility
Neck Shim (2 Reg. Rods)

Fig. 11.7. ATR core layout. Source: D. R. DeBoisblanc,
3rd ICPUAE, 1964.

The most advanced research reactor core is that developed for the High Flux
Isotope Reactor (HFIR) at Oak Ridge, which was designed primarily as a flux
trap reactor in which to produce transuranic elements for research purposes.
While this has been a major task of HFIR, the reactor has also been used for
neutron research. The reactor core consists of four concentric regions
(Fig. 11.9). The fuelled region consists of two series of fuel plates that
are held between concentric aluminium cylinders. The inner series contains
171 plates containing a total of 2.6 kg of ^{235}U and 2.12 g of ^{10}B burnable
poison. The outer series contains 369 plates loaded with 6.8 kg of ^{235}U and
no burnable poison. The total assembly of 9.4 kg of ^{235}U constitutes the
core.

Each fuel plate is of sandwich construction with a cermet of 35 wt % U_3O_8
dispersed in 65 wt % Al in the inner plates and 40 wt % U_3O_8 in the outer
plates (Fig. 11.10). The plates are 0.050 in thick, with a cladding thickness
0.010 in. They have an involute curve which provides a coolant channel of
constant width 0.050 in between plates. Note in Fig. 11.10 the varying shape
of the fuel "meat" and the $^{10}B_4C$ dispersion in the inner plates.

The reactor has a maximum power of 100 MW, giving a peak unperturbed thermal
neutron flux of 5 x 10^{15} n/cm^2/sec. A typical fuel cycle is 14 days. The
maximum power density is 4.3 MW/ℓ, peak heat flux = 2.0 x 10^6 Btu/hr ft^2.
Typical coolant conditions are 600 psig with a peak clad temperature of 627°F
(331°C). Oxide build-up on the cladding causes the hot spot temperature to
rise 200°F during the fuel cycle.

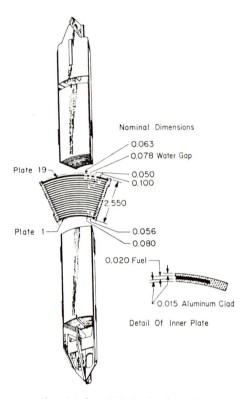

Nominal Dimensions

0.063
0.078 Water Gap

Plate 19

0.050
0.100

2.550

Plate 1

0.056
0.080

0.020 Fuel

0.015 Aluminum Clad

Detail Of Inner Plate

Fig. 11.8.   ATR fuel element.

As noted above, plate-type dispersion fuel elements may fail by a variety of processes. If the fuel core is not properly bonded to the cladding, blisters may form due to gas pressure build-up in the unbonded region. Poor design giving inadequate or nonuniform cooling may lead to warping or buckling. However, the "inherent" limit on performance is due to fuel swelling. A theory of dispersion fuels has been described by Weir [7]. Each fissile particle creates a severely damaged region around it due to fission fragment recoil, as shown in Fig. 11.11. The width of the zone is ∿10 μm. The swelling of each fuel particle is restrained by a web of less damaged metal outside the recoil zone. Thus we can portray an ideal dispersion fuel as a regular array of fuel spheres of diameter D, each surrounded by a recoil zone of width λ_m (Fig. 11.12a). The separation between the particles or the width of the less damaged matrix, d', depends on the fuel fraction V_f and the particle diameter D, according to the plot in Fig. 11.12b. One must strive to maximize the value of the parameter d', since this represents the load-bearing part of the dispersion fuel. In practice, dispersions are far from ideal and failure may occur as the recoil zones crack and link up to produce a plane of failure (Fig. 11.13). Cracking is exacerbated by the neutron radiation hardening of the metal matrix with a consequent loss of ductility.

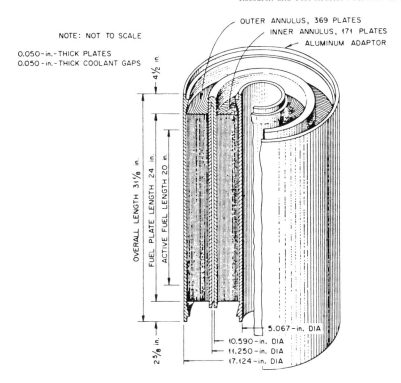

NOTE: NOT TO SCALE

0.050-in.-THICK PLATES
0.050-in.-THICK COOLANT GAPS

OUTER ANNULUS, 369 PLATES
INNER ANNULUS, 171 PLATES
ALUMINUM ADAPTOR

OVERALL LENGTH 31⅛ in.
FUEL PLATE LENGTH 24 in.
ACTIVE FUEL LENGTH 20 in.
4½ in.
2⅝ in.

5.067-in. DIA
10.590-in. DIA
11.250-in. DIA
17.124-in. DIA

Fig. 11.9. HFIR fuel element. Source: J. W. Swartout
et al., 3rd ICPUAE, 1964.

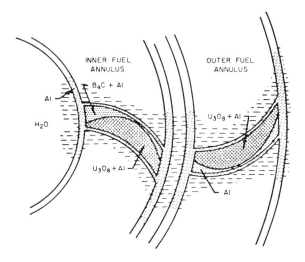

INNER FUEL
ANNULUS

OUTER FUEL
ANNULUS

B₄C + Al
Al
H₂O
U₃O₈ + Al
U₃O₈ + Al
Al

Fig. 11.10. Schematic representation of core cross
section, showing fuel contours. Source:
J. W. Swartout *et al.*

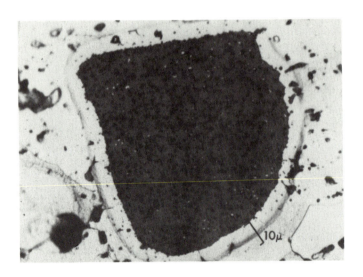

Fig. 11.11.   Optical micrograph showing recoil damage
zone around UO_2 particle approximately 900X.
Source: B. Frost and M. Waldron.

As mentioned above, a typical end-of-life burn-up for research reactor
dispersion fuel is 25-30% of ^{235}U atoms. This is determined by trial and
error rather than by scientific method or by modelling.

Studies of research reactor dispersion fuels revived in the late 1970s with
realization that many countries possessed fully enriched uranium in the form
of fuel elements for research and test reactors. It would be comparatively
simple to extract that fissile material in order to make a nuclear weapon.
Consequently, attempts have been made to reduce the enrichment level
(preferably to 20% or less) by increasing the uranium loading of the
dispersion fuel. Moderate success has been obtained by increasing the UAl_3
and U_3O_8 fractions in the MTR and HFIR fuel plates, respectively. However,
more useful gains have been made by dispersing the compund U_3Si in aluminium.

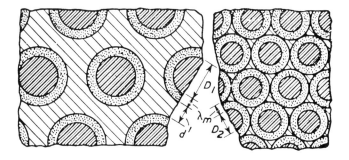

(a) FOR CONSTANT VOLUME FRACTION OF PARTICLE PHASE

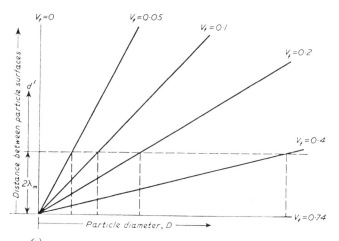

(b) FOR VARIOUS VOLUME FRACTIONS, V_f, OF PARTICLE PHASE

Fig. 11.12.   Effect of particle size and spacing on
undamaged distance between particles.

U_3Si has a density of ~ 15 g/cm^3 or a very much higher uranium density than
UAl_3. Dispersions of up to 45 v/o U_3Si in aluminium have been fabricated
into plate form and successfully irradiated in the ORR reactor[8]. These
elements begin to approach the target of 20% enrichment. Even if nonproliferation policy changes, the work will be technologically useful, since these
fuel designs with higher uranium enrichments will permit the upgrading of
existing cores, provided the coolant can handle the heat loading.

A totally different research reactor concept is the TRIGA developed by General
Atomics Corporation in the 1950s [9]. This water-cooled (and partly moderated)
reactor was designed to be inherently safe. The fuel element contains a
uranium-zirconium-hydride fuel consisting of 8 wt % uranium of 20% enrichment,
91 wt % zirconium and 1 wt % hydrogen, clad in 1100 aluminium. The fuel is
made by first alloying uranium and zirconium and then hydriding [10]. Not

only is the fuel resistant to radiation effects, it also has a negative coefficient of thermal expansion due to hydrogen redistribution from hot to cold regions, giving the latter larger dimensions [11]. This buils inherent safety into the fuel and TRIGA reactors have regularly been sent prompt critical to give a burst of neutrons which is very rapidly damped by the negative feedback effect; the prompt neutron lifetime is $\sim 1.5 \times 10^{-4}$ sec.

Different again from the TRIGA reactor is TREAT, a pulsed graphite moderated reactor built by Argonne at INEL for the transient testing of fast reactor fuel elements [12].

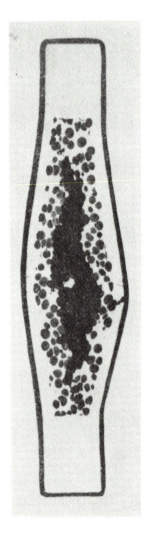

Fig. 11.13.   Failure in a 50 vol. % cermet after 13.5 % burn-up at 650°C.   Source: B. Frost and M. Waldron.

A somewhat similar reactor CABRI has been built at Cadarache in France. The TREAT fuel element is illustrated in Fig. 11.14. The fuel is a matrix of porous, reactor grade graphite in which U_3O_8 is dispersed at a level of 0.248 wt %. The fuel is clad in Zircaloy sheets welded to form a box shape 4 in × 4 in × 72 in (10 cm × 10 cm × 183 cm) long, with a 48 in (122 cm) fuelled section.

The fine dispersion of the fuel in the moderator is important for a pulsed reactor in which one is relying on a rapid negative reactivity coefficient in the graphite to quench the pulse. The pulse of 1000 MW-sec is initiated by withdrawing control rod elements to drive the reactor prompt critical. The pulse levels off due to the reactivity feedback effect and control rods are reinserted. The reactor is cooled over a long period by flowing air. The maximum graphite temperature during a pulse is $400^{\circ}C$ and the average is $260^{\circ}C$. The peak neutron flux is 10^{16} nv and the average over the core is 3.4×10^{15} nvt.

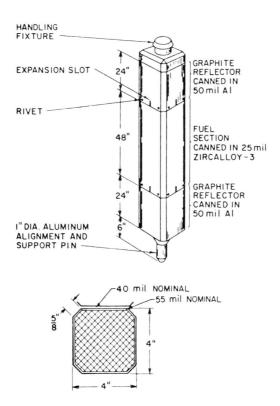

Fig. 11.14. TREAT fuel element. Fuel: uranium oxide in graphite. <u>Source</u>: Argonne National Laboratory.

The TREAT core is being upgraded by construction of longer elements containing a dispersion of UC_2 in graphite that can operate to higher temperatures. The upgraded core will permit the pulsing of full size fast reactor subassemblies, typically ∿8 ft (200 cm) long with a 3 ft (92 cm) fuel section.

Finally, we will briefly discuss the ultimate in pulsed neutron sources — an accelerator driven spallation source, typified by Argonne's IPNS-1 (Intense Pulsed Neutron Source, No. 1) [13]. An alternating synchrotron produces a beam of protons at 500 MeV at 8-20 μA current and in pulses at a frequency of 30-45 Hz. The beam strikes a uranium target where spallation reactions occur that produce many neutrons per pulse. Figure 11.15 shows the distribution of neutrons in a uranium target at 5×10^{12} protons/pulse at 30 Hz. Heat removal from the target is important to minimize thermal cycling growth or tearing effects [14]. The uranium is of the "adjusted" composition discussed earlier. It is vacuum melted and cast and then machined into discs 4 in (10 cm) diameter and 1.08 in (2.7 cm) thick. They are clad in Zircaloy-2 which is sealed by electron beam welding and finally hot-isostatically bonded at 15,000 psi in pure helium at $840^\circ C$ for 4 hours to produce total bonding of the uranium to the cladding. The finished product is shown in Fig. 11.16.

The most likely failure mechanism is fatigue failure of the cladding due to shutdowns and start-ups (not to the "normal" pulsing). Some simple thermal cycling experiments that simulated start-up and shutdown cycles led to an estimated life of 5000 cycles. One expects about 10 cycles per day during operation, hence a life of 500 days or at least 2 years (and more likely 3 years) of operation before failure.

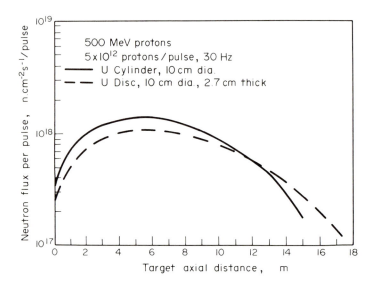

Fig. 11.15.   IPNS target neutron yield.   Source: Argonne
National Laboratory.

Fig. 11.16.   Zircaloy-2 clad uranium target for IPNS.
Source: Argonne National Laboratory.

REFERENCES

1. Boyle, E. J. and Cunningham, J. E., *Proceedings of the First International Conference on the Peaceful Uses of Atomic Energy*, Vol. 9, p. 953, UN (1956).
2. Hendrie, J. M., *Proceedings of the Third International Conference on the Peaceful Uses of Atomic Energy*, Paper P/222, UN (1965).
3. Shuck, A., Haugen, J., Isserow, S. and Huber, R., Specification and Procurement of CP-5 Fuel Tubes, Argonne National Laboratory Report ANL-7708 (1970).
4. Dopchie, H. and Planquart, J., *Proceedings of the Second International Conference on the Peaceful Uses of Atomic Energy*, Paper P/1679, UN (1958).
5. deBoisblanc, D. R., Gordon, R. H., Lazar, A. H. and Weber, L. J., *Proceedings of the Third International Conference on the Peaceful Uses of Atomic Energy*, Paper P/223, UN (1965).
6. Beaver, R. J., *et al.*, USAEC Report ORNL-3632 (1964).
7. Weir, J. R., USAEC Report ORNL-2902 (1960).
8. Domagala, R. F., Wiencek, T. C., Thresh, H. R. and Stahl, D., paper to International Meeting on the Development, Fabrication and Application of Reduced-Enrichment Fuels for Research and Test Reactors, at Argonne National Laboratory (November 1980).
9. Koutz, S. L., *et al.*, *Proceedings of the Second International Conference on the Peaceful Uses of Atomic Energy*, Paper P/1017, UN (1958).
10. Merten, U., *et al.*, ibid, Paper P.789.
11. Lillie, A. F., *et al.*, USAEC Report AI-AEC-13084 (1973).
12. Freund, G. A., Iskenderian, H. P. and Okrent, D., *Proceedings of the Second International Conference on the Peaceful Uses of Atomic Energy*, Paper P/1848, UN (1958).
13. Carpenter, J. A., *Nucl. Instrum. & Methods*, 145, 91 (1977).
14. Loomis, B. A., Thresh, H. R., Fogle, G. L. and Gerber, S. B., *Nuclear Technology* (in the press).

CHAPTER 12

Unconventional Fuel Elements

In the 1970s nuclear power became technically conservative and after TMI it has become ultra-conservative. INFCE, NASAP and CONAES[*] have debated the nuclear options and have generally concluded that the world needs nuclear power, but that no new approaches are needed, or are indeed desirable. INFCE in particular reviewed other options, including alternative fuel cycles such as the thorium cycle in CANDUs, Light Water Breeders and HTGRs, dual cycles (starting the fuel in a LWR and transferring it to a HWR) and other expedients to minimize the risks of nuclear proliferation. However, no system emerged as superior to those currently under development.

The field of nuclear energy was not always so conservative, and many novel systems have been considered, some being taken to the test or demonstration reactor scale. Lest we forget the lessons of the fifties and sixties and in case new opportunities arise in the future, this chapter will discuss unusual concepts.

The following quote from Freeman Dyson's book[**] appropriately describes the transition in the nuclear industry:

> "The fundamental problem of the nuclear power industry is not reactor safety, not waste disposal, not the dangers of nuclear proliferation, real though all these problems are. The fundamental problem of the industry is that nobody any longer has any fun building reactors. It is inconceivable under present conditions that a group of enthusiasts could assemble in a schoolhouse and design, build, test, license and sell a reactor[†] within three years. Sometime between 1960 and 1970, the fun went out of the business. The adventurers, the experimenters, the inventors, were driven out, and the accountants and managers took control.

---

[*]INFCE, International Nuclear Fuel Cycle Evaluation: NASAP, Nonproliferation Alternative Systems Assessment Program; CONAES, Committee on Nuclear and Alternative Energy Systems (National Academy of Sciences).

[**]Freeman Dyson, *Disturbing the Universe*, pp.104-105, Harper & Row, New York (1979).

[†]GA's TRIGA reactor.

Not only in private industry but also in the government laboratories, at Los Alamos, Livermore, Oak Ridge and Argonnne, the groups of bright young people who used to build and invent and experiment with a great variety of reactors were disbanded. The accountants and managers decided that it was not cost effective to let bright people play with weird reactors. So the weird reactors disappeared and with them the chance of any radical improvement beyond our existing systems. We are left with a very small number of reactor types in operation, each of them frozen into a huge bureaucratic organization that makes any substantial change impossible, each of them in various ways technically unsatisfactory, each of them less safe than many possible alternative designs which have been discarded. Nobody builds reactors for fun any more. The spirit of the little red schoolhouse is dead. That, in my opinion, is what went wrong with nuclear power."

We can approach this discussion in several ways. One is to consider new reactor concepts from the standpoint of the different branches of nuclear technology. Thus, in seeking more thermally efficient, cheaper, more resource-efficient, special purpose or even more interesting reactors, the reactor physicist thinks in terms of such concepts as mixed spectrum reactors, spectral shift, mixed composition cores (e.g. the "parfait" LMFBR), cosine loading, burnable poisons, and alternative moderators (such as beryllium and metal hydrides).

The engineer will think in terms of alternative cooling or energy-conversion systems such as fog or dust cooling, liquid metal boiling (e.g. potassium vapour turbine), MHD, thermionic (direct) conversion, topping and bottoming cycles and reactors for process heat or congeneration.

The materials scientist or nuclear metallurgist is concerned with materials integrity and hence how to minimize mechanical stresses or to restrain the consequences of fission. This could include the venting of fuel elements, modification of metal and ceramic fuels by "alloying", the examination of "new" (i.e. known but untested) compounds, novel dispersion fuels and liquid fuels including metallic, aqueous and molten salt solutions and two-phase suspensions or slurries.

Of course, reactor development involves a combination of physics, engineering and metallurgy for every system, but in the past an advocate from one branch has often led the way. For example, the nuclear chemists at Oak Ridge led the development of the homogeneous aqueous reactor to the stage of a reactor experiment. Unfortunately the concept foundered on a very difficult corrosion problem: fission events close to the normally protective oxide film on metallic containers led to an early breakdown and rapid corrosion.

Another approach is from the fuel element viewpoint. For each new system thought up by the engineer or physicist, the metallurgist has to examine his inventory of fuel materials for the most suitable fuel and cladding materials. If we discount fluid fuels for the moment, the inventory of fuels may be described as follows:

Metal	U, Pu, Th and combinations
Alloys	U-Mo, U-Zr, U-Pu-Zr, etc.
Intermetallic Compounds	UAl_3, U_3Si, UBe_{13}, etc.
Ceramics	UO_2, UC, UN, US, UP, etc.
Dispersions or Two-phase Fuels	
Alloy	UAl_{13} in Al
	UB_{13} in Be
Oxide-metal	UO_2 in stainless steel
Oxide-oxide	UO_2 in BeO
Carbide-graphite	UC_2 in C
Hydrides	UH_3 in ZrH_2

As with the engineering- and physics-oriented systems described above, most of these systems have been fairly thoroughly tested. New applications (which are rare today) brought new challenges. Thermionic reactor cores required fuel elements to operate with surface temperatures around $1800^\circ C$, so that a dispersion of a stable fuel (UO_2) in a very high melting point metal (Mo or W) with suitable electron emitting characteristics had to be examined. For the nuclear rocket a stable, high-temperature fuel was needed; candidates were (again) UO_2 in tungsten and UC_2 dispersed uncoated in graphite. For the aircaft nuclear propulsion programme special cladding materials were developed to resist high-temperature oxidation.

Fuel element geometry is another area where conservatism is dominant, i.e. the cylindrical pin is the standard, with flat plates as the second most popular design. An interesting deviation from convention is the "inside-out" design where one virbatorily compacts powdered UO_2 into a subassembly duct through which cylindrical coolant tubes pass. Then the duct assumes the role of the primary barrier to fission product and fuel escape. The cylindrical tubes are very strong in compression against the swelling fuel. Similarly, a European GCFR design used SiC-coated particles held in a wire mesh container which was cooled by helium. The AVR pebble-bed HTGR is not unlike this, both systems having the advantage of continuous or on-load fuel changing under gravity.

In the area of reactor safety some ingenious ideas have been proposed, but not often adopted. The original version of the DFR fuel element consisted of an annular casting of U-Mo fuel with a niobium outer tube and a vanadium inner tube. The fuel reacts with vanadium at a lower temperature than with molybdenum, hence a reactor transient would result in a meltdown occurring down the inside hole — much preferable to it occurring down the outside.

A fusible link at the top end of a subassembly is another clever idea for a fast reactor core. In the event of a transient, the low-melting-point link would melt and the subassembly would drop out of the core, reducing the reactivity. A varient is for the fusible link to be a uranium alloy that heats up to its melting point as the neutron flux rises above its normal level.

Ingenuity extends to the fabrication of fuels. A group at Harwell developed a centrifugal casting process for making carbide "shot" or spheres. The fuel rod is made one anode of an arc melter and either it or the hearth is spun, the molten fuel flying off as molten spheres that solidify before landing. The process is expensive, but has the potential advantage of acting as a reprocessing step in which irradiated fuel is melted and the volatile fission products are removed, reminiscent of the EBR-II integral reprocessing of metal fuel.

New powder metallurgy processes, such as hot isostatic pressuring (HIP) and the direct rolling of powders to sheet, give the metallurgist new techniques for making fuels more efficiently or in different shapes. The HIP method, in which a "green" object is placed in a furnace and subjected to very high gas pressures, not only processes to near final shape but is an excellent way to bond fuel to cladding. The uranium targets of the Argonne Intense Pulsed Neutron Source are finished by hot isostatically bonding cylinders of cast depleted and adjusted uranium to electron beam welded Zircaloy cladding, thus ensuring a good bond for heat transfer.

Earlier in this chapter reference was made to the Homogeneous Aqueous Reactor. It represents a class of fluid fuel reactors in which an attempt was made to escape from the constraints of the solid rod-type fuel element. Basically, the reactor core consists of a solution or suspension of uranium or plutonium in a convenient solvent. In the case of thermal reactors the solvent may be water, in which case it acts as the moderator, or there may be solid moderator throughout the core. Advantages of fluid fuel reactors are the absence of fuel swelling, improved neutron economy and on-line reprocessing; i.e. continuous removal of gaseous and other fission products and fuel make-up. The three principal contenders, well described in the book by Lane, McPherson and Maslan,[*] are:

1. Homogeneous Aqueous Reactors, which are generally solutions of uranyl sulphate or nitrate in H_2O or D_2O.
2. Molten-salt Reactors, in which the fuel is most commonly a mixture of the fluorides of uranium, alkali metals and beryllium or zirconium, with melting points in the range 850–950°F.
3. Liquid Metal Fuel Reactors, the most studied of which uses a solution of uranium in bismuth as the fuel. A fast reactor design that used a liquid metal fuel was called LAMPRE (Los Alamos Molten Plutonium Reactor Experiment). This used the plutonium-iron eutectic, melting at 410°C, as the fuel. The design was not that of a fluid fuel reactor, but used tantalum tubes to contain the fuel; these were cooled with liquid sodium as in conventional fast reactors.

The principal problem in all of these systems was corrosion. Much effort was expended in finding containers that would remain intact for times in excess of 2 years of operation. Each project claimed that they had solved this problem, but by that stage funding had been discontinued as a result of decisions to focus on the more conventional core designs.

Given the complexity of fuel element development today and the somewhat limited resources being applied to this area, it is very reasonable to concentrate on improving the well-tried design rather than to develop new ones. The main purpose of this brief chapter has been to remind the reader of the breadth of choice that exists in fuel element designs and the fact that many of these options have been examined in the past.

---

[*]*Fluid Fuel Reactors*, Addison-Wesley (1958).

Some Useful General References

1. Glasstone, S. and Sesonske, A., *Nuclear Reactor Engineering*, Van Nostrand-Reinhold, Princeton, NJ (1967).
2. Sesonske, A., *Nuclear Power Plant Design Analysis*, TID-26241, NTIS.
3. Benedict, M. and Pigford, T., *Nuclear Chemical Engineering*, McGraw-Hill, New York (1976).
4. Simnad, M. T. and Howe, J. P., *Materials Science in Engineering Technology*, Chapter 2, Academic Press (1979).
5. Simnad, M. T. *Fuel Element Experience in Nuclear Power Reactors*, ANS-AEC Monograph, Gordon & Breach, New York (1971).
6. Nero, Anthony V., *A Guidebook to Nuclear Reactors*, University of California Press, Berkeley (1979).
7. American Physical Society special reports, published in *Reviews of Modern Physics*:
   Radiation Effects on Materials (1975)
   Light Water Reactor Safety (1975)
   Nuclear Fuel Cycles and Waste Management (1977)
8. Fuel Elements, Materials and Reactor Volumes of the *Proceedings of UN Conferences on the Peaceful Uses of Atomic Energy*: 1955, 1958, 1964, and 1971.
9. *Directory of Nuclear Reactors*, IAEA, Vienna. Ten volumes from ∿1958 through 1976.
10. *Nuclear Research Index*, Francis Hodgson, Guernsey, U.K., 5th Edition (1976).
11. *The Atomic Energy Deskbook*, Ed. John F. Hogerton, Reinhold, New York (1963).

Tabulation of Industrial Capabilities in the USA

Capability	Agns	Allied	A.I.	B&W	(B&W) Numec (Apollo)	(B&W) Numec (Leechburg)	B.M.I.	CE. (Hematite)	CE. (Windsor)	ENC	GAC (San Diego)	GAC (Youngsville)	GAC (Site)	G.E. (Morris)	G.E. (San Jose)	G.E. (Wilmington)	K.Mc (Cimarron)	K.Mc (Sequoyah)
$U_3O_8 \rightarrow UF_6$		X																X
$UF_6 \rightarrow UO_2$			F	X			X			X					X		X	
UO_2 pellets				X	X		X	X		X					X		X	
UO_2 fuel fabr.*				X	X			X		X					X		X	
Carbide fuels			X		X						X	F						
Special fuels			X		X						X	F					X	
U-233 fuels					X								F					
Thorium					X						X	F					X	
Pu fuels			X	X	X				F	X					X		X	
U fuel R&D			X	X	X		X	X		X	X				X		X	
Pu fuel R&D			X		X		X	X		X	X				X		X	
Depl. U-metal			X		X												X	
Depl. U-cpds			X		X												X	
Cold U scrap					X					F					X		X	X
Cold Pu scrap		F			X					F					X			
Spent fuel	F									F			F	F				
Sl. enr. $U \rightarrow UF_6$	F																	
Hi. enr. $U \rightarrow UF_6$					X													
Location of Facility	Barnwell, S.C.	Metropolis, Ill.	Canoga Park, Calif.	Lynchburg, Va.	Apollo, Penn.	Leechburg, Penn. (Park Township)	Columbus, Ohio	Hematite, Mo.	Windsor, Conn.	Richland, Wash.	San Diego, Calif.	Youngsville, N.C.	Site not yet announced	Morris, Ill.	San Jose and Vallecitos, Calif.	Wilmington, N.C.	Cimarron, Okla.	Sequoyah, Okla.
Name of Company	Allied General Nuclear Services	Allied Chemical Corporation	Atomics International Division Rockwell International	The Babcock & Wilcox Company	NUMEC A Subsidiary of the Babcock & Wilcox Company		Battelle Memorial Institute	Combustion Engineering, Inc.		Exxon Nuclear Co. An Affiliate of Exxon Corporation	General Atomic Company			General Electric Company			Kerr McGee Nuclear Corporation	

X denotes present domestic capability;

N.L.	N.F.S.	NM	TNS	TEXAS	U.N.C.	U.S.N.	W	Capability	
								$U_3O_8 \rightarrow UF_6$	Conversion of ore concentrates to UF_6
	X						X	$UF_6 \rightarrow UO_2$	Conversion of enriched UF_6 to UO_2
	X						X	UO_2 pellets	Production of UO_2 pellets from UO_2 powder
							X	UO_2 fuel fabr.*	Fabrication of fuel elements containing UO_2 pellets*
	X					X		Carbide fuels	Proc. and/or fabr. of high temp. fuels, i.e. carbides, coated particles, etc.
		X		X	X			Special fuels	Proc. and/or fabr. of special fuels, e.g. research reactor. U-A1, HFIR, EBR-11, CP-5, etc.
	X							U-233 fuels	Proc. and/or fabr. of fuels containing U-233
	X		X					Thorium	Proc. and/or fabr. of core components containing thorium
	X						F X	Pu fuels	Proc. and/or fabr. of fuels containing plutonium
	X						X	U fuel R&D	Research and devel. on uranium fuels
							F X	Pu fuel R&D	Research and devel. on plutonium fuels
X	X	X	X					Depl. U-metal	Fabr. of speciality metal parts from depleted uranium
	X		X					Depl. U-cpds	Prodn. of special compounds containing depleted uranium
	X				X		F	Cold U scrap	Processing of scrap containing unirradiated uranium
	X						F X	Cold Pu scrap	Processing of scrap containing unirradiated plutonium
	F							Spent fuel	Reprocessing of irradiated fuel
	F							Sl.enr. $U \rightarrow UF_6$	Conversion to UF_6 of uranium containing less than 5.0% U-235 (from oxides and UNH)
								Hi.enr. $U \rightarrow UF_6$	Conversion to UF_6 of uranium containing more than 5.0% U-235 (from oxides and UNH)

Locations:
- N.L.: Erwin, Tenn.? — NL Industries
- N.F.S.: Erwin, Tenn. — Nuclear Fuel Services Inc. (Owned by Getty Skelly oil companies)
- West Valley, N.Y.
- NM: West Concord, Mass. — Nuclear Metals Div. Whittaker Corp.
- TNS: Jonesboro, Tenn. — Tennessee Nuclear Specialties, Inc.
- TEXAS: Attleboro, Mass. — * Texas Instruments Incorporated
- U.N.C.: Wood River Jct., R.I. — United Nuclear Corporation
- U.S.N.: Oak Ridge, Tenn. — U.S. Nuclear
- Anderson, S.C.
- Cheswick, Penn. — Westinghouse Electric Corp.
- Columbia, S.C.

*Not limited to companies that are competing for commercial power reactor reloads, but includes some companies with limited capacity.

U.S. Atomic Energy Commission
Office of Industry Relations
December 1974

Source: WASH-1174-74

F denotes future domestic capability.   * Ceased operations in 1981.

Index

Absorption cross sections 78
Acceptance criteria for ECCS 197, 198
Adjusted uranium 21, 201
Advanced gas-cooled reactor (AGR) 8, 200, 206-210
Advanced test reactor (ATR) 254
Alpha-Gamma hot cell facility 159
Amoeba effect 213, 216
Apparent diffusion coefficient 64

Ballooning 196, 197
BEMOD code 121
Boiling water reactor (BWR) 181
BOR-60 232
BR-2 250
BR-5 reactor 224
Breakaway corrosion 81
Breeding 217
BUBL code 65
Burn-up 75, 76

CABRI 240
CANDU 8, 24, 184
Carbides 13, 236, 237
Cermets 35, 72, 232
Characterization 40
Cladding 7
Closed loop in-reactor assembly (CLIRA) 155, 156
Coated particle fuels 36, 210-216
COMETHE code 122, 124, 127
Commercialization 140
Coolable geometry 197
CP-5 reactor 249-250

Core (of reactor) 3
Corrosion of Zircaloy 85

Defect clusters 89
Defects 88
Densification 60, 186
Departure from Nucleate Boiling (DNB) 247
Derby 15
DFR 225-227
Direct electrical heating (DEH) 133-135
Dispersion fuels 23, 72
Displacements per atom (dpa) 95
Ducts 79

EBR-I reactor 219-224
EBR-II reactor 136, 227
Emergency core cooling systems (ECCS) 193, 195
Engineering test reactor (ETR)
Enrichment 15

Fabrication 132, 152
Failure (of fuel elements) 6
Failure mechanisms 184
Fast burst 202
Fast flux test facility (FFTF) 137, 235
Fast mixed spectrum reactor 242
Fermi reactor 227
Fertile materials 13
Fission 1
Fission cross section 50

273

Fission fragments 1
Fission gas release 55, 62-67
Fission gases 54
Fission products (f.p.) 1, 8, 51
Fission track 50
Fissium 21, 227
French gas cooled reactors 204
Fretting and wear 186
Fuel-cladding interactions 86, 234
Fuel cycle 16, 17
Fuel cycle cost 17
Fuel element codes 104
Fuel element models 104
Fuel properties 34

Gap conductance 103
Gas-cooled fast reactor (GCFR) 100
Gloveboxes 159-151
Graphite 211
GRASS code 66
GRASS-SST 122, 123, 128
Growth (of uranium) 20

Heat generation rate 100
Heavy water reactor (HWR) 181
HFEF 159
High Activity Handling Building
    (Harwell) 162
High flux beam reactor (HFBR) 248
High flux isotope reactor (HFIR) 255
High temperature gas-cooled reactor
    (HTGR) 36, 210-216
Hot cell 158-176
Hot isostatic pressing (HIP) 268

Intense pulsed neutron source (IPNS)
    262, 263
Interstitial 50
Irradiation 20
Iodine 7
Ion simulation (of neutron damage)
    95

LAMPRE 268
Licensing of cores 190-192
LIFE code 105, 116-121
Liquid fuels 23, 268
Liquid Metal Fast Breeder Reactor
    (LMFBR) 98
Liquid metal fuel reactors 268
LOCA 67, 192
LOFT 193

Magnox 80, 201
Materials test reactor (MTR) 248
Metal fuels 18

National Energy Software Center 116
Nondestructive examination (NDE) 138
Nonproliferation 241

Oxygen : metal ratio 25, 43, 53

Pellet-cladding interaction (PCI) 7,
    70, 140, 186
PFR 235, 236
Phase diagrams 41-47
Phase rule 41
Phenix 236
Picture frame 24
Plate-type elements 248
Plutonium 1, 7, 13, 42
Postirradiation examination 158
Posttest examination 138
Pressurized water reactor (PWR) 3,
    177-180
Proliferation resistance 10
Pumping power 102

Quality assurance 141

Radiation damage 52
Radiation effects 52
Radiation-induced segregation 93
Rapsodie 236
Refuelling 245
Resolution 57
Restructuring (of UO_2) 60
Rod cluster control (RCC) 177
Roughened cladding 79
Run-beyond-cladding-breach (RBCB)
    238

Separate effects tests 10
Sipping 176
Sol-gel process 37
Stainless steels 83-85
Subassembly 3
SWELL 105, 113-116
Swelling 52, 55

Test capsule 152

Test vehicle 152
Thermal gradients 102, 103
Thermal conductivity (UO$_2$) 27, 102, 103
Thermal cycling 20
Thorium 23
TREAT 240, 260-261
TRIGA 259, 260

UNCLE code 121
Uraninite 14
Uranium 1, 13-16, 201
Uranium dioxide (UO$_2$) 15-24
U-Al 43
U-Mo-Ti 49
U-Nb-Zr 48
U-Pu-Zr 22

U-Zr 25
UC (U,Pu)C 31
UN 33
(U,Pu)O$_2$ 29
U$_3$Si 259

Vacancy 50
Vipac fuel 233
Void swelling 90

Wastage allowance 78
Weibull statistics 137-138

Zircaloys 82
Zirconium 80, 85
Zirconium hydride 81, 184